AF882564

GROUNDWORK

ISBN 978-0-615-50107-9
Library of Congress Control Number 2011910195

Gordon Stillman Photography
Richmond, Virginia
www.gordonstillman.com

Virginia Sustainable Agriculture Documentary Project
Richmond, Virginia
www.vasadp.org

Book Design by Lauren O'Neill | Design

a documentary photography project by **GORDON STILLMAN**

GROUNDWORK

I would like to thank everyone who contributed a pledge to help publish *Groundwork*.
I would especially like to thank:

The TeamWork Group, Inc.
Anne S. Nordeman
Joshua and Jamie Eftekhari
My parents, Adelaide Villmoare and Peter Stillman

I would also like to thank Neah Monteiro for her support and guidance throughout the project.

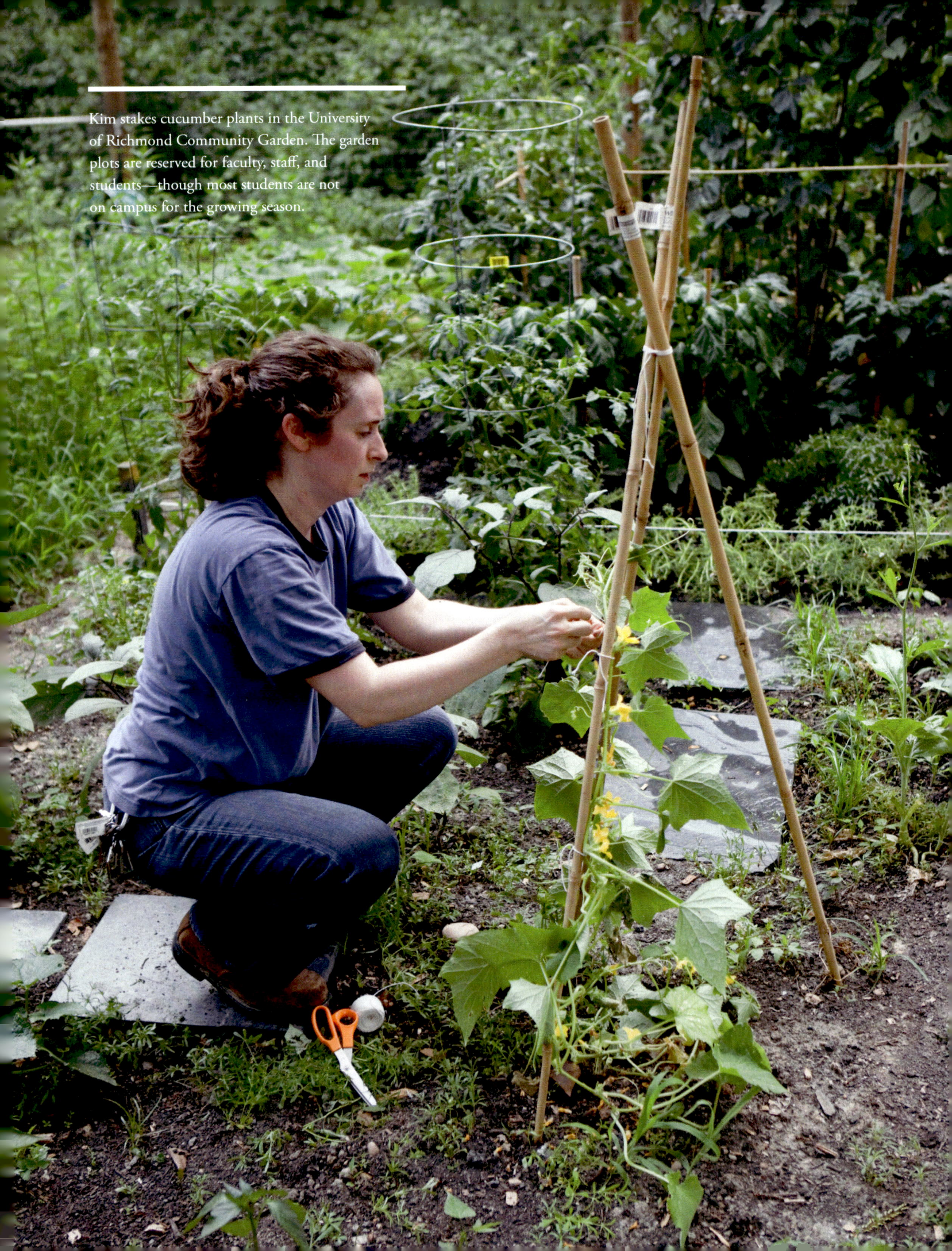

Kim stakes cucumber plants in the University of Richmond Community Garden. The garden plots are reserved for faculty, staff, and students—though most students are not on campus for the growing season.

GORDON STILLMAN

Groundwork **is the culmination of two years of photographing small farms, families, markets, restaurants, community gardens and many others involved with local food in Virginia. I began the work as the Virginia Sustainable Agriculture Documentary Project (VASADP) and still host a journal with each shoot at www.vasadp.org. I began photographing because I was interested in the future of food production and access and our health. I am interested in how we produce food and how we can change food production to be more healthy, equitable and environmentally friendly.**

Agricultural practices have changed radically over the past 100 years—sometimes with unintended consequences. These practices include the use of synthetic fertilizers, pesticides, heavy machinery and genetically modified crops. All of these technologies have allowed for a huge increase in yield, but they have also brought unintended consequences, for example, pollution in rivers and fisheries from runoff and from pesticides.

All of these advances also mean that fewer people need to farm, and combined with global trade, our food can come from anywhere around the world. Many people have lost touch with where their food actually comes from, how it is produced and its effects on our health and the environment. At the same time, many people are now getting back in touch with where their food comes from and the impacts of their food through local farmers markets, co-ops, buying programs and organic sections in grocery stores.

Groundwork provides a view of local food in Virginia, not only to show where our food comes from, but also to show sustainable practices for growing, distributing and eating food. The local food movement provides a vibrant example of individuals, families and communities who care about food, and who see it as a primary vessel for reducing human impact on the environment and for enriching communities.

The local food movement has exploded in Virginia's capital city, Richmond, in the last four years. The main way people in Richmond can purchase local food is at farmers markets and through community supported agriculture programs (CSAs), where consumers purchase a share of the farm's produce over a season and receive a weekly delivery. CSAs provide a very intimate link between the consumer and the food they eat: Subscribers know the farmers, and many farmers hold events at their farms so members can see how their food is grown. Most farmers that sell at markets also maintain CSA programs.

Victory Farms in Hanover County has a CSA day in the summer, where everyone brings a dish, preferably one made with ingredients from the farm. Members get to see how the farm works and meet all the farmers and other CSA members. Agriberry CSA invites its members to the farm on occasion to see how their berries are grown and get some hands-on experience to see how their food is produced from seed to market.

Markets and CSA programs allow everyone to see the sources of their food, and reintroduce people to how their food is grown—so they know more about their food than just from which grocery store they bought it.

This relationship with what you eat—and who grew it—is a strong building block for sustainable food production. It is one that makes you think about food beyond the nutrients it provides. On a local level, food becomes as much about sustaining your local farmer and environment as it is about sustaining you. And this can lead to new and innovative thinking to create a completely sustainable food system.

Tricycle Gardens and Farm To Family are two organizations trying to create a completely sustainable food economy—one in which all food is sustainable for the grower, the distributor and the community. This requires not only that food be sustainably produced, but also economically viable for the farmer to grow and for families to buy, and abundant enough for everyone to consume.

This relationship with what you eat—and who grew it—is a strong building block for sustainable food production.

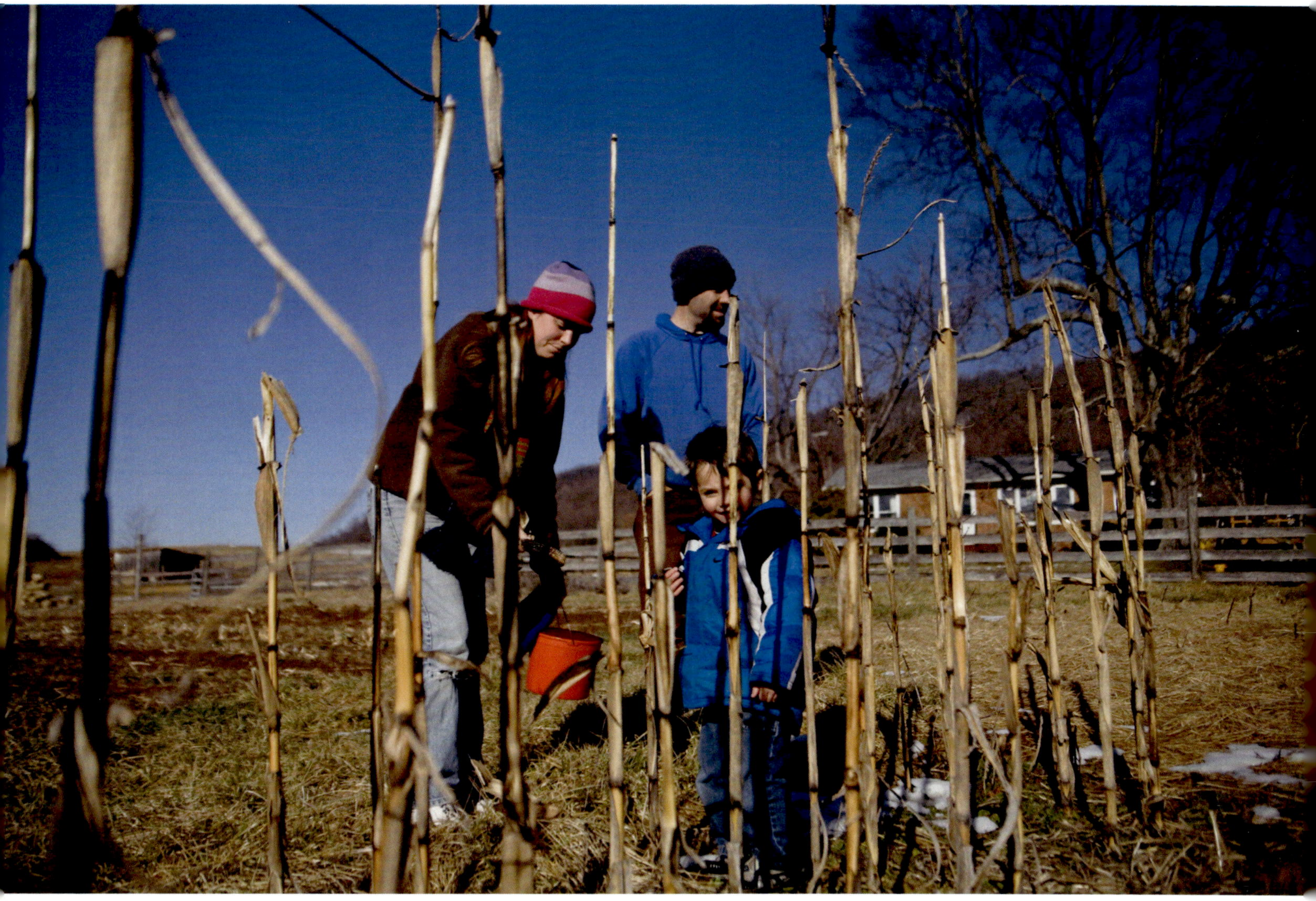

Michael, Kathryn, and Eli survey their farm, Appalachia Star Farm, in early spring. They planted their fall corn crop because Eli wanted colorful corn for the holidays.

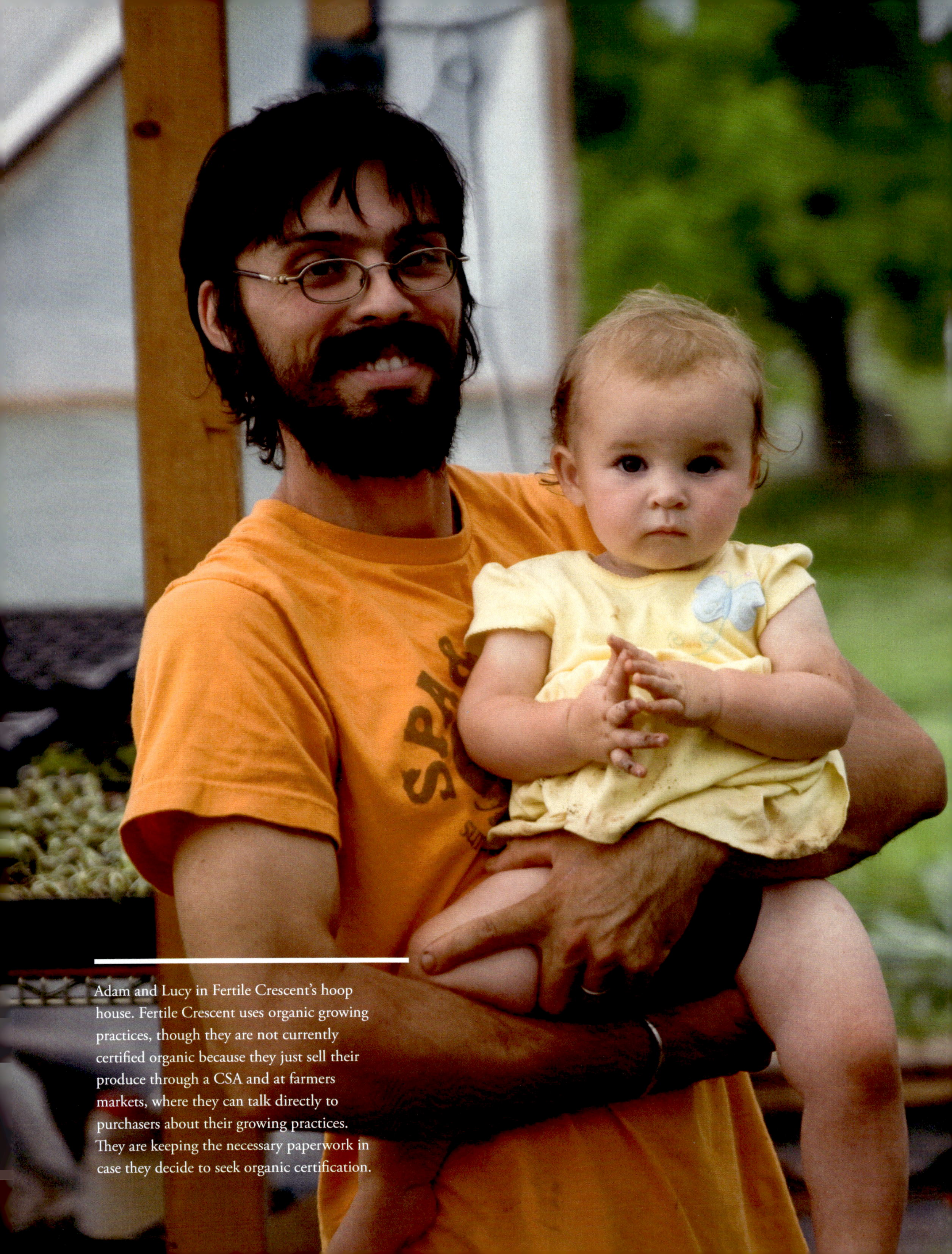

Adam and Lucy in Fertile Crescent's hoop house. Fertile Crescent uses organic growing practices, though they are not currently certified organic because they just sell their produce through a CSA and at farmers markets, where they can talk directly to purchasers about their growing practices. They are keeping the necessary paperwork in case they decide to seek organic certification.

Tricycle Gardens operates an urban farm, community gardens and food education programs. The organization is constantly innovating and trying to make food more available, affordable and enjoyable. Their programs seek to feed the community good food, but also to create avenues for everyone to have sustainable and local food, and they are based on building a better understanding of our food and how the entire process, from growing to eating, affects us.

While Tricycle Gardens brings food knowledge and growing into Richmond, Farm To Family creates a distribution network for local food year-round. They operate a store, greenhouse and mobile market on a converted school bus. The benefits from this network include a more viable means for many farmers to sell their produce and also the ability, with the mobile bus market, to reach communities that do not have access to fresh food. Creating such alternative distribution networks—ones that can reach any community and service local growers—is another important aspect in developing new ways to make food sustainable.

Groundwork looks at many local innovations that help create a sustainable food economy. Some practices, like backyard gardens, have been popular for generations, and other practices are relatively new, like mobile markets and urban farms. Making such sustainable practices viable and fitting them into the larger food economy is still a work in progress—and the ultimate goal for many. This may require involvement with grocers from Whole Foods to Walmart and others in the larger farming industry—the Tysons, Smithfields and Green Giants. But before many larger companies get fully involved, it is up to those working at the local level to experiment with sustainable practices to prove what works and what can be adopted on a national and global scale.

Gordon Stillman is a professional photographer who started the Virginia Sustainable Agriculture Documentary Project as a personal project in 2009.

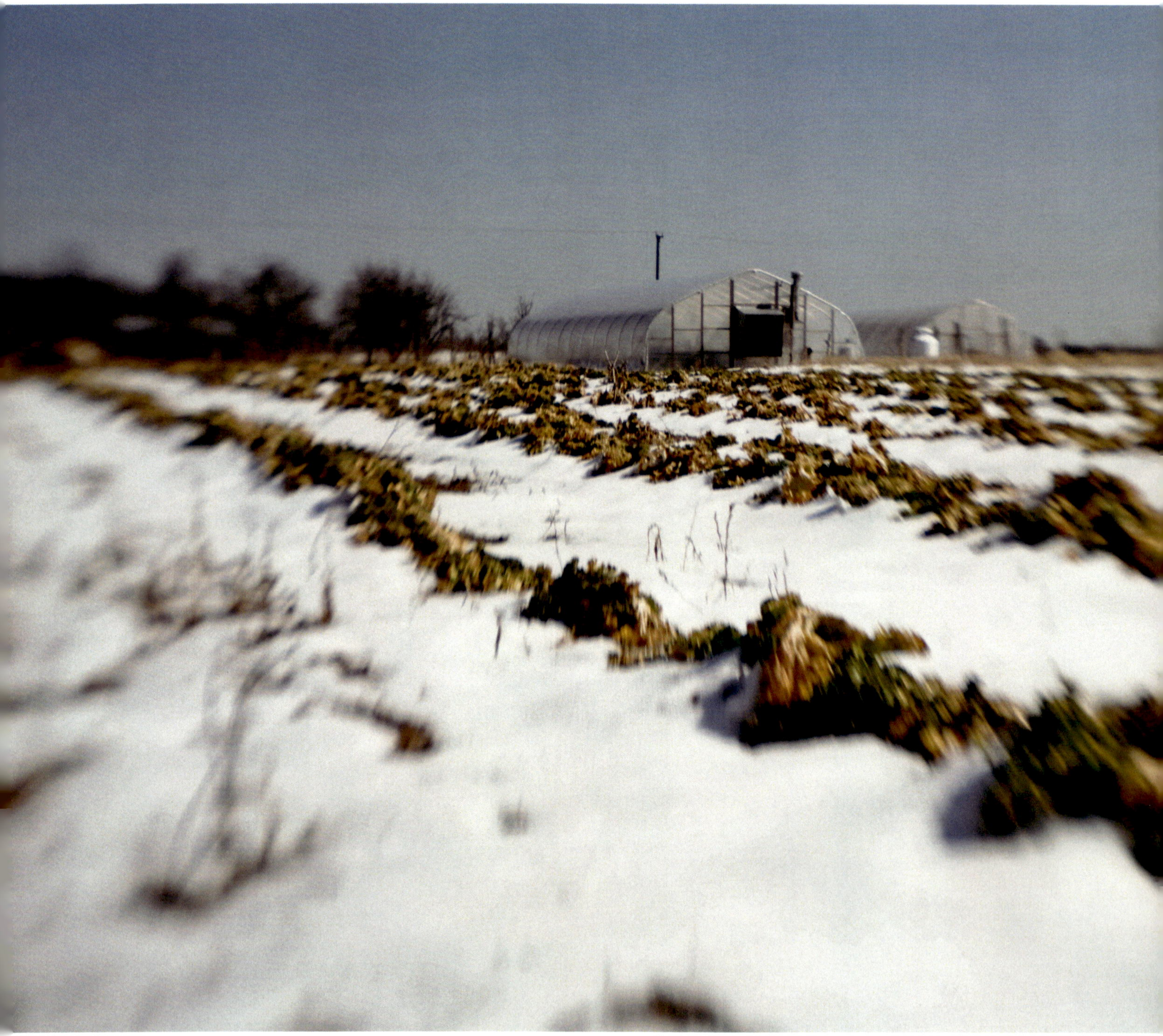

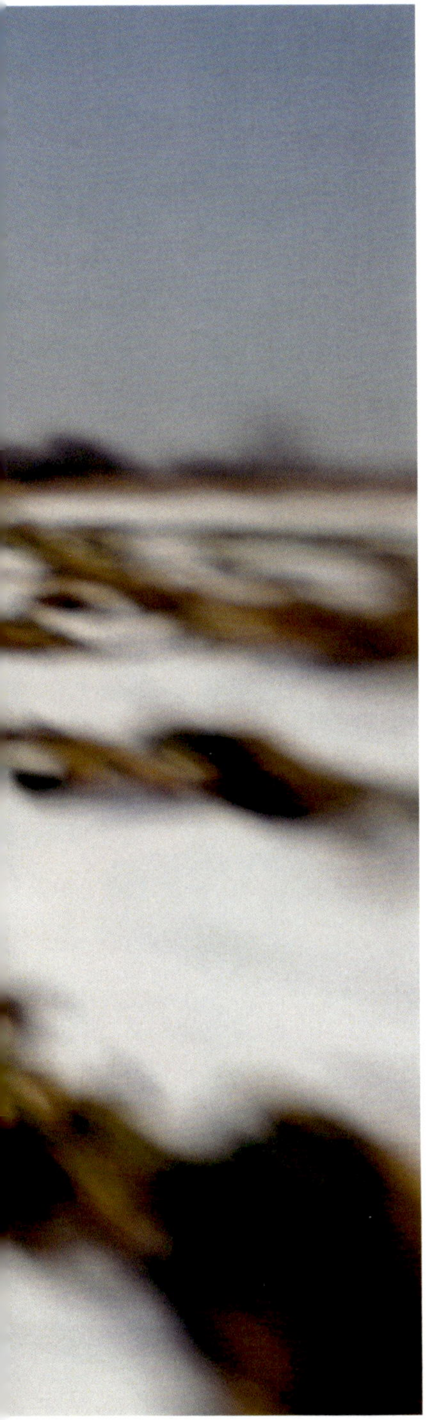

Snowfall in early spring delays planting and tilling at Dayspring Farm in Cologne, Virginia. Farming is still dependent on the weather; snow or too much or too little rain can delay planting and harvesting, as it did in 2009. Every year presents new challenges that make it hard to predict the farm's crops and farmers' income. Because of this, many farms operate community supported agriculture (CSA) programs where members of the community pay a seasonal fee and then receive weekly shares of produce. CSAs provide income for farms in the early spring when it is most needed. They also put some of the risk onto the shareholder—if the farm produces less, their share is smaller.

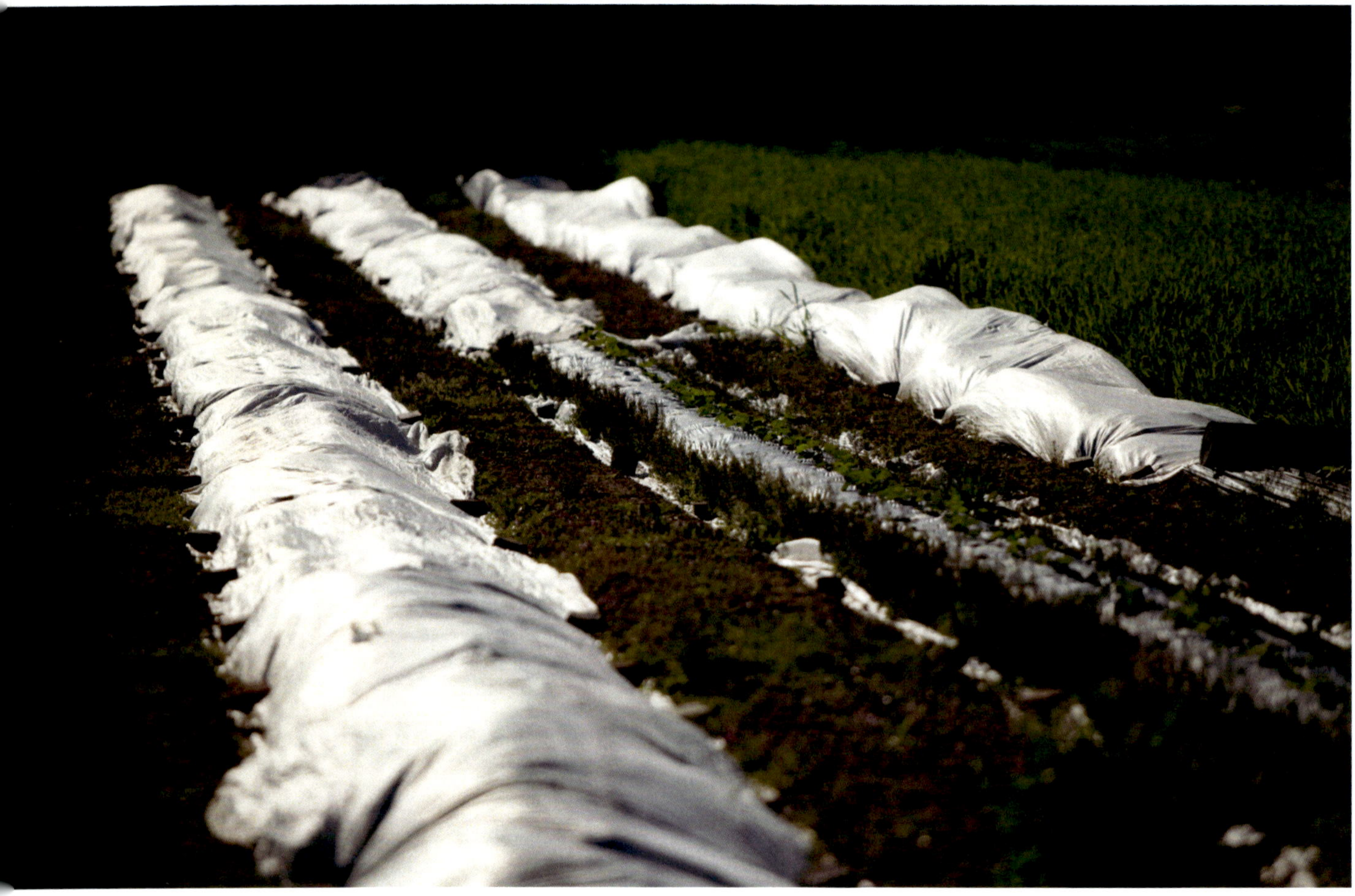

Crop blankets protect plants—strawberries in this case—from cold weather and pests, including bugs and deer. The blankets not only extend the growing season but provide a pesticide-free way to discourage pests.

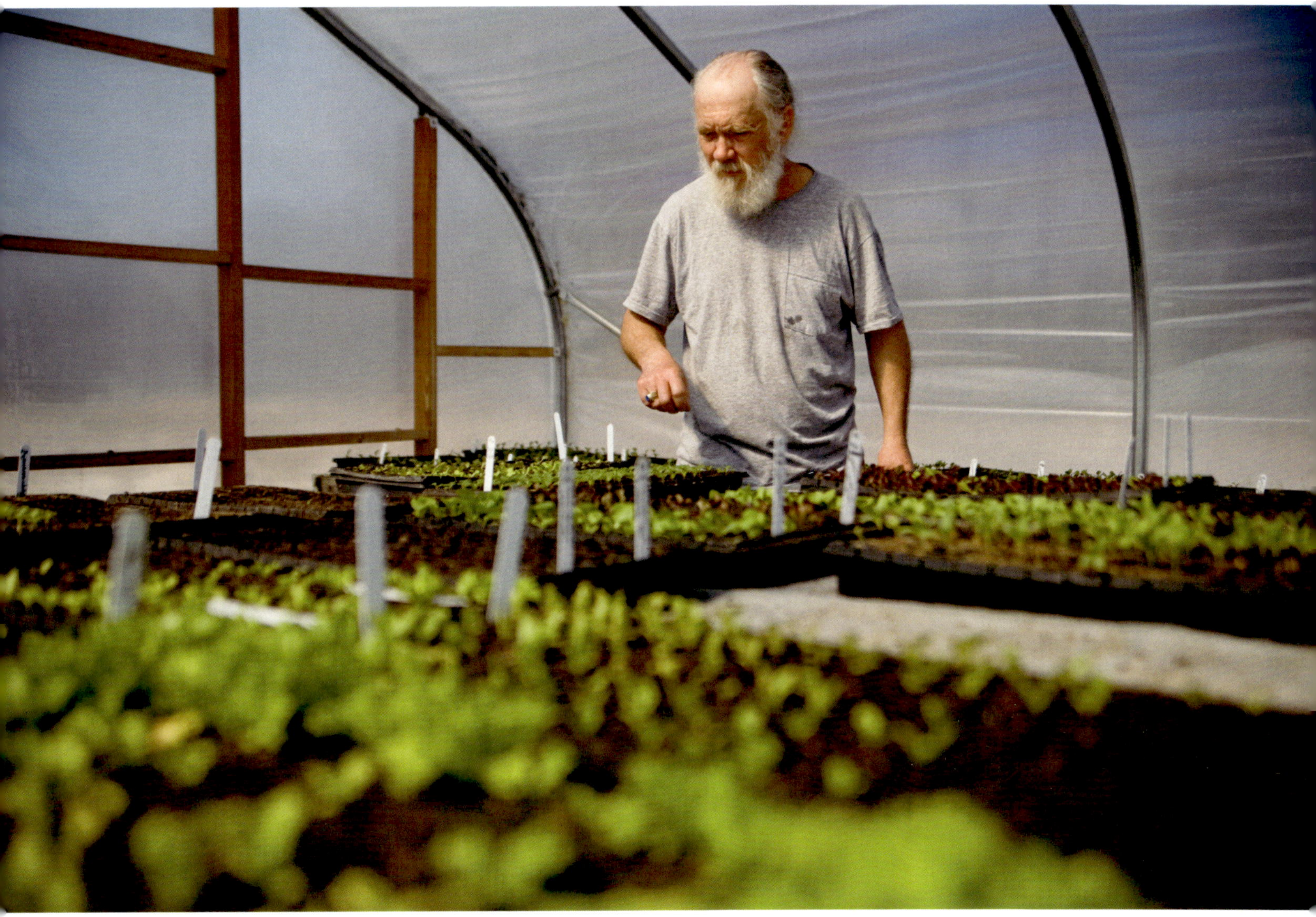

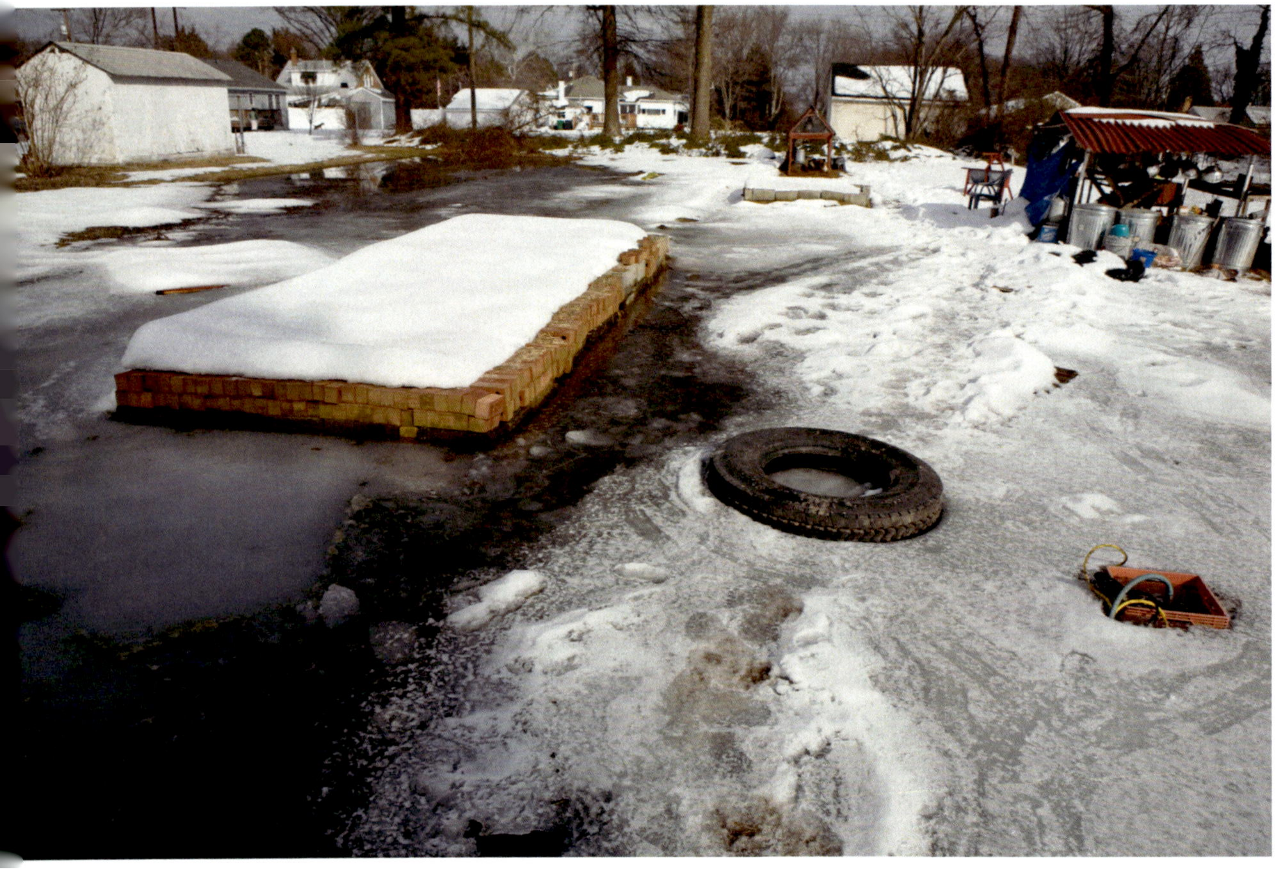

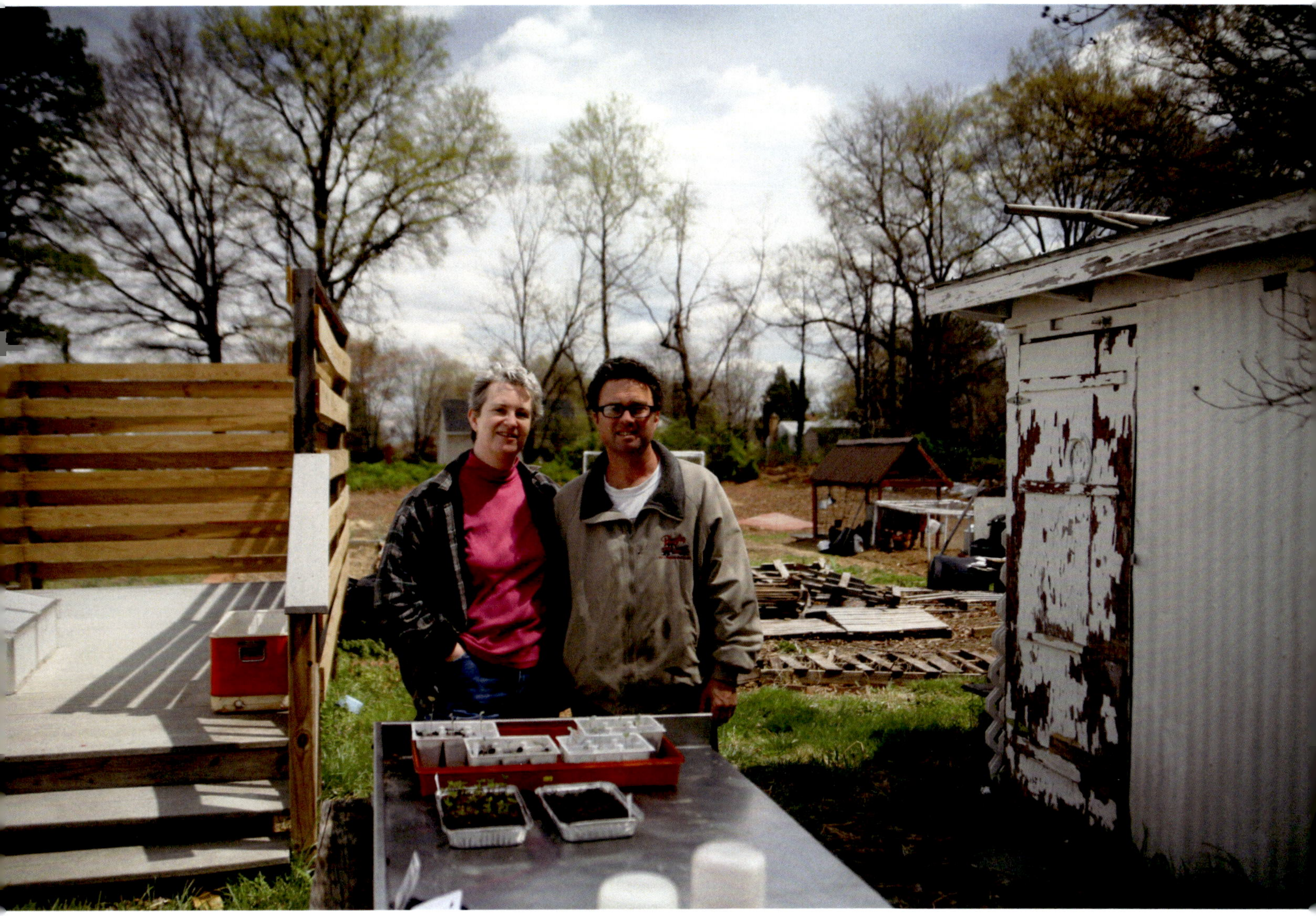

In a snow-covered backyard garden in Richmond, the asparagus patch is surrounded by brick, and chickens are in the hen house. Rachel and Lee started gardening to save money on groceries and have since expanded their garden to provide most of their food during the summer. They preserve or can their harvests to last throughout the winter.

Chickens at Appalachia Star Farm provide more than eggs, they also help fertilize and control pests—many farms have chicken trucks or mobile hen houses so they can let the chickens slowly move over the entire farm.

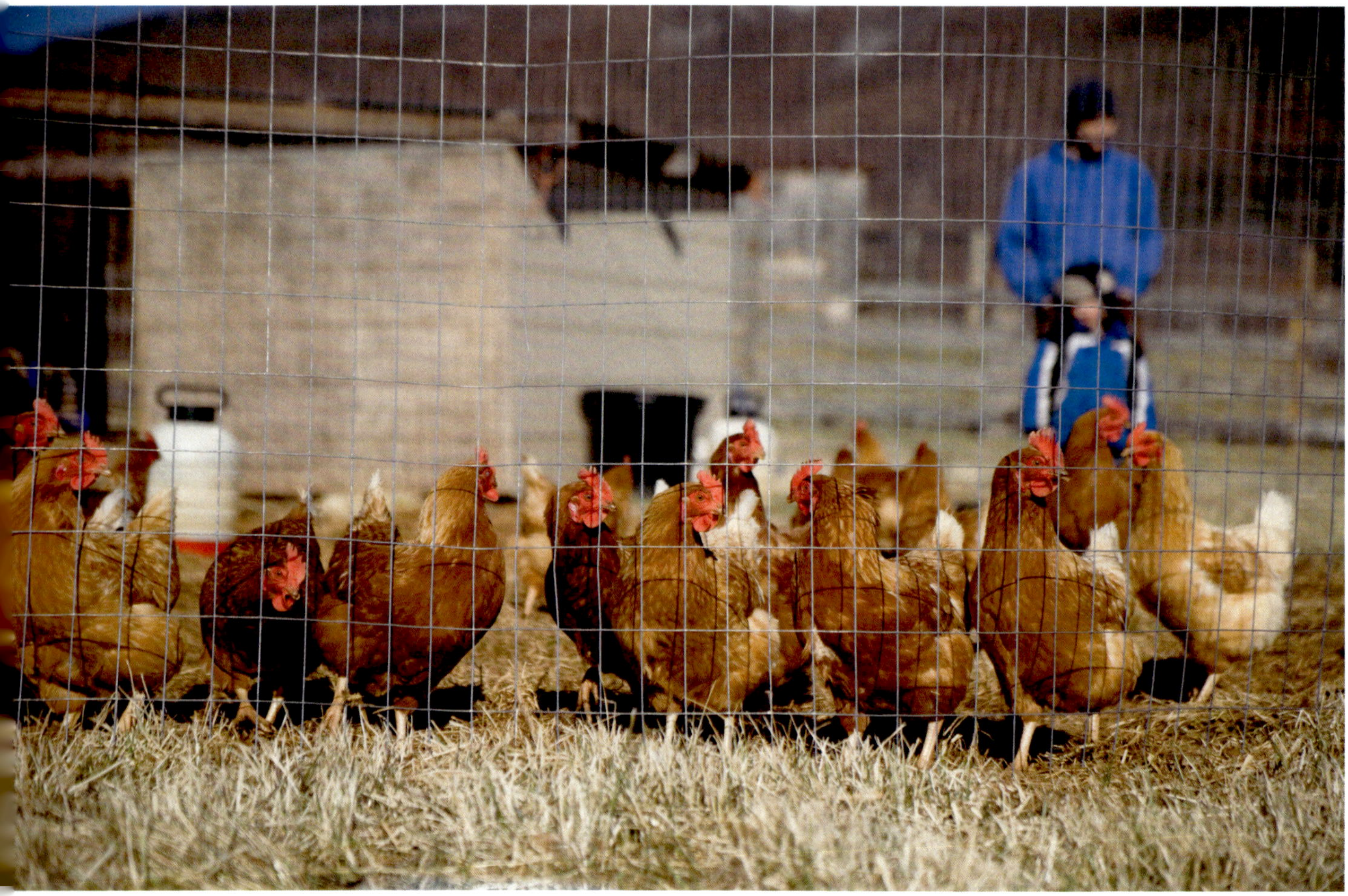

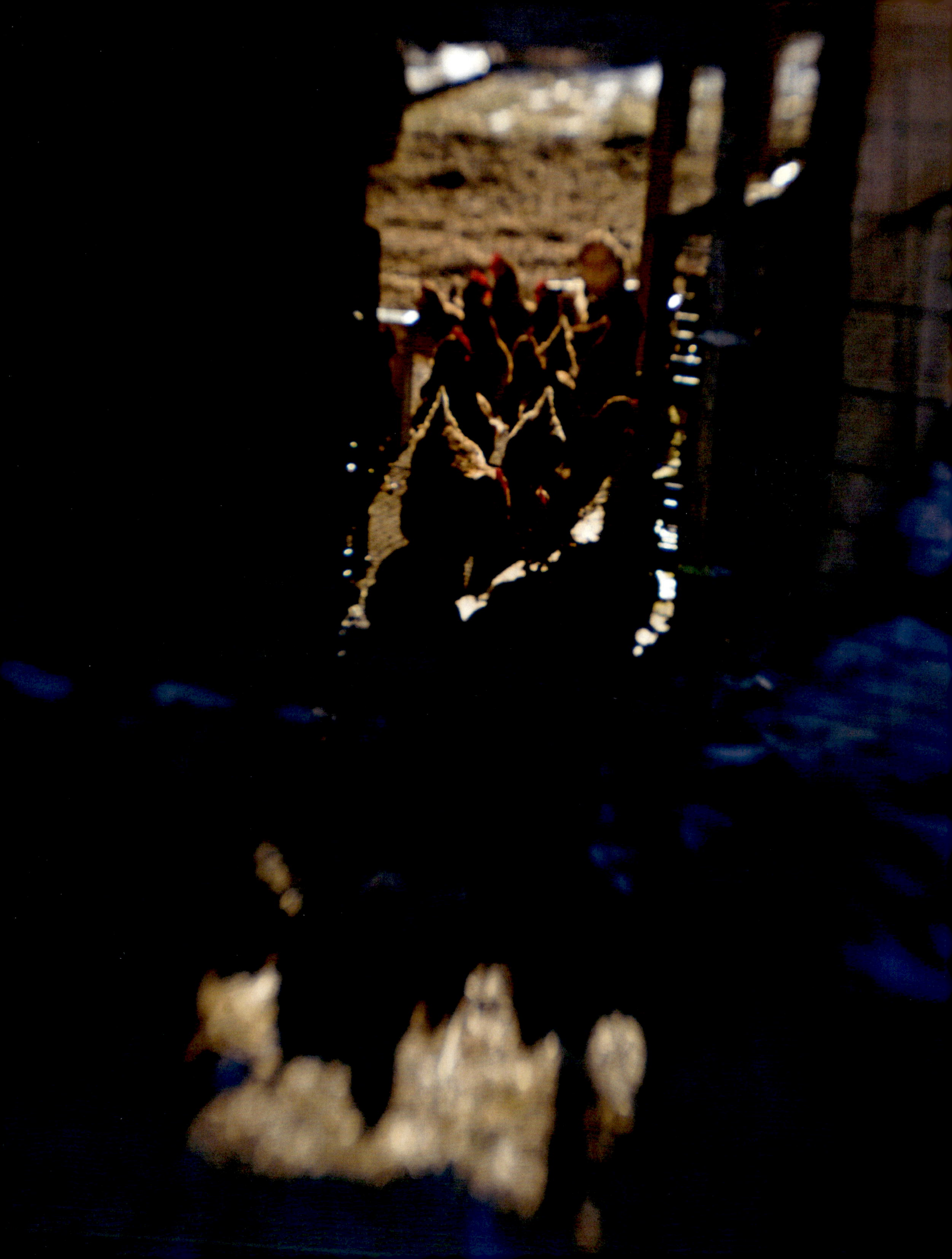

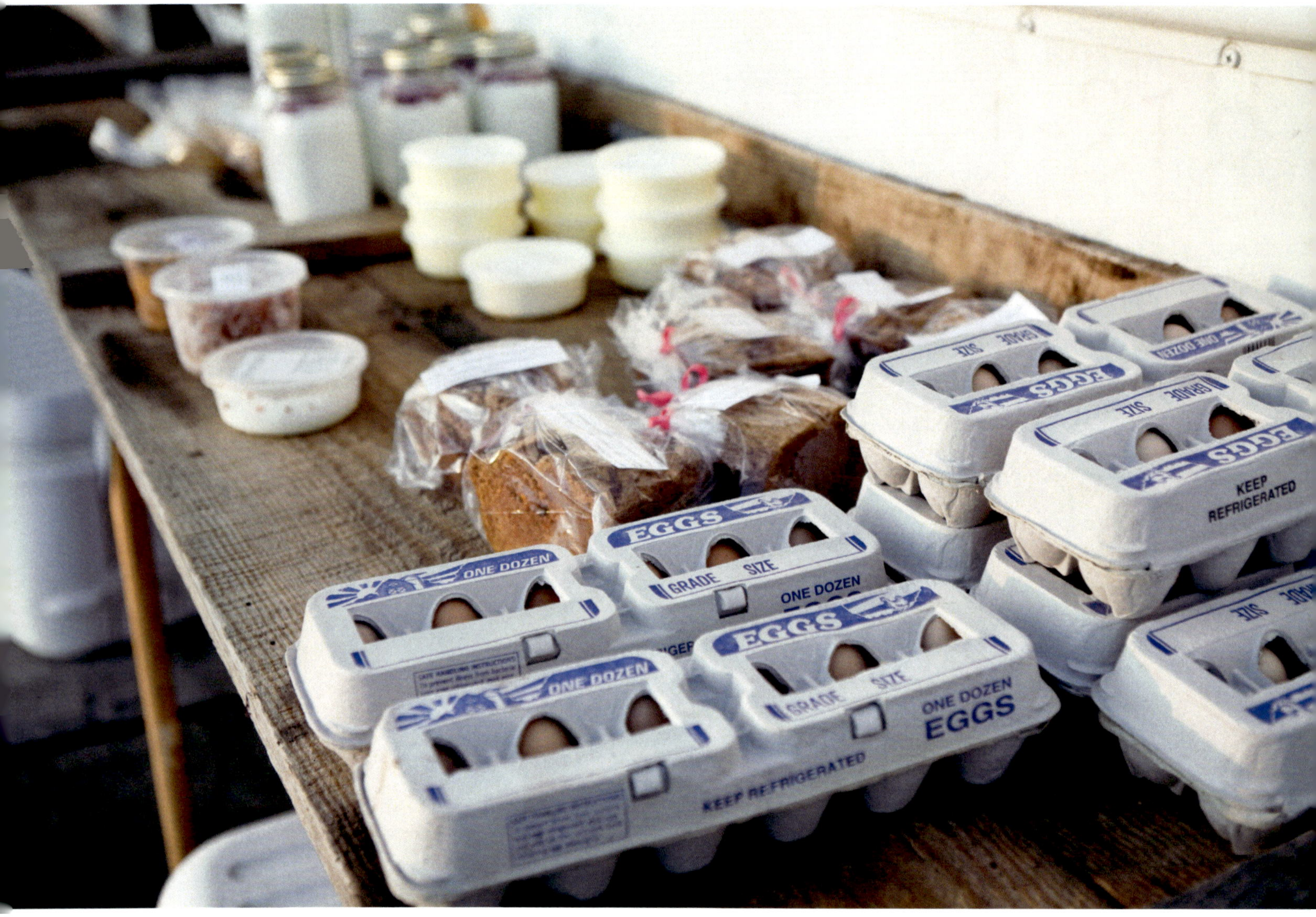

Offerings from Farm To Family's meat and dairy CSA include eggs and Mountain View butter, cheese and yogurt. Farm To Family's CSA is available all year, and they have established numerous points throughout Richmond where shareholders can conveniently pick up their shares off the bus.

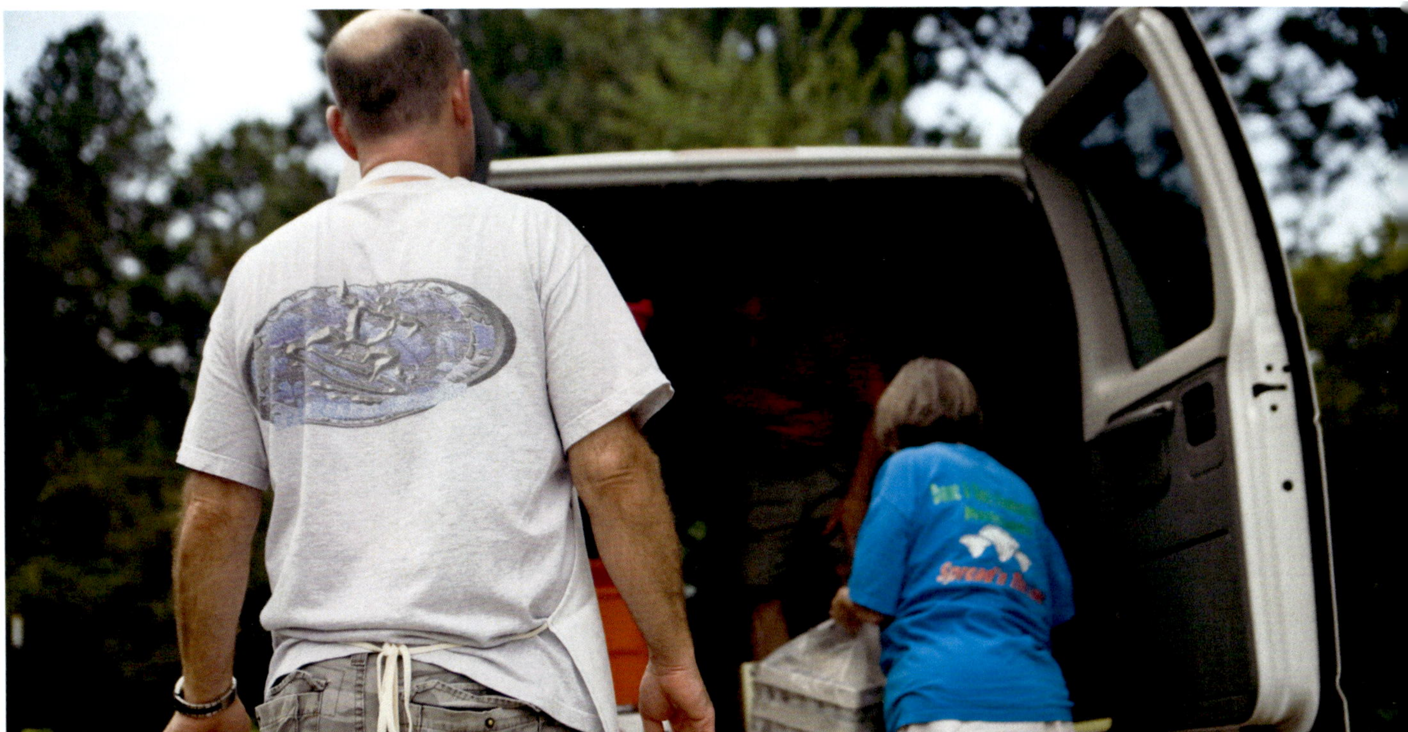

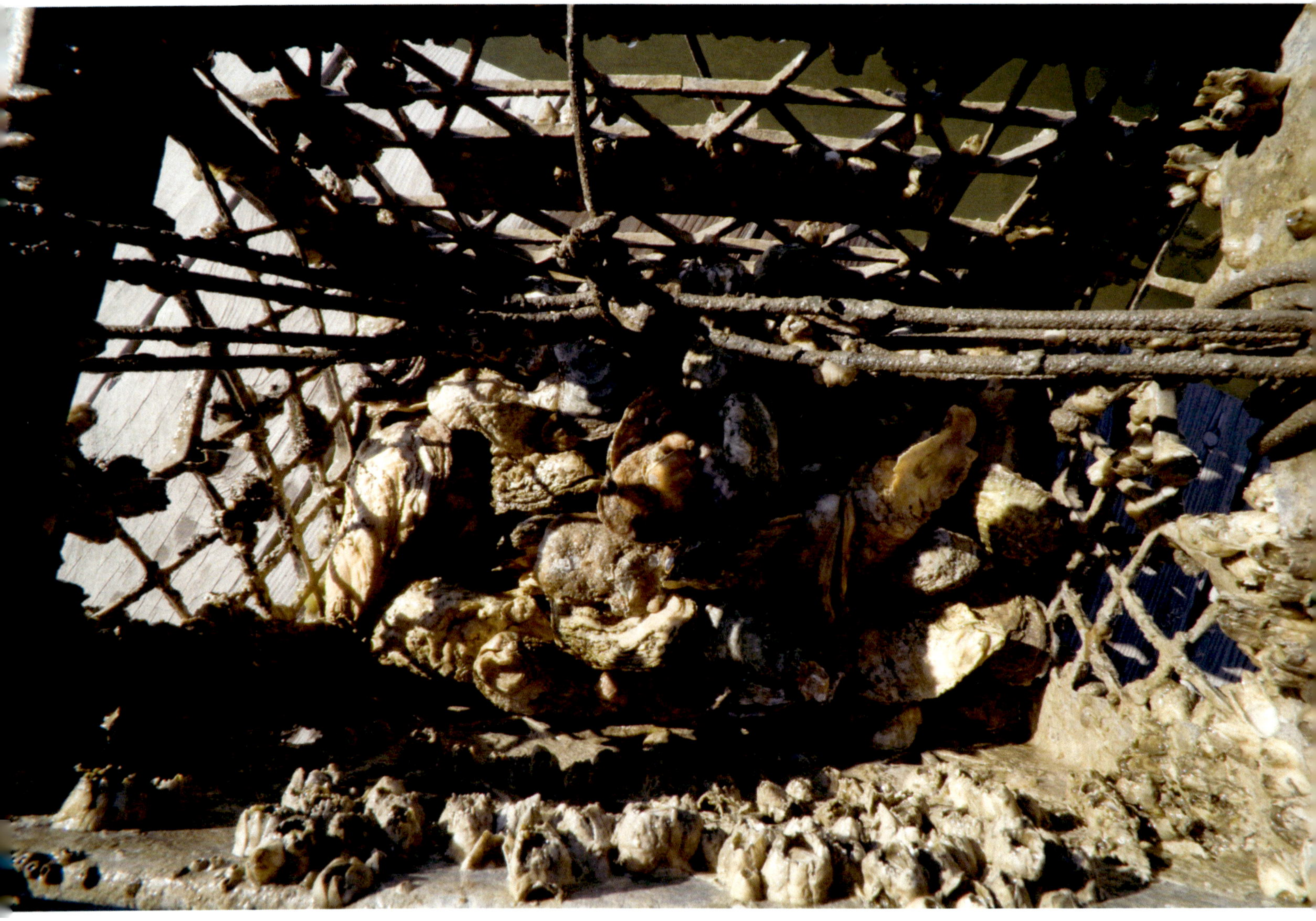

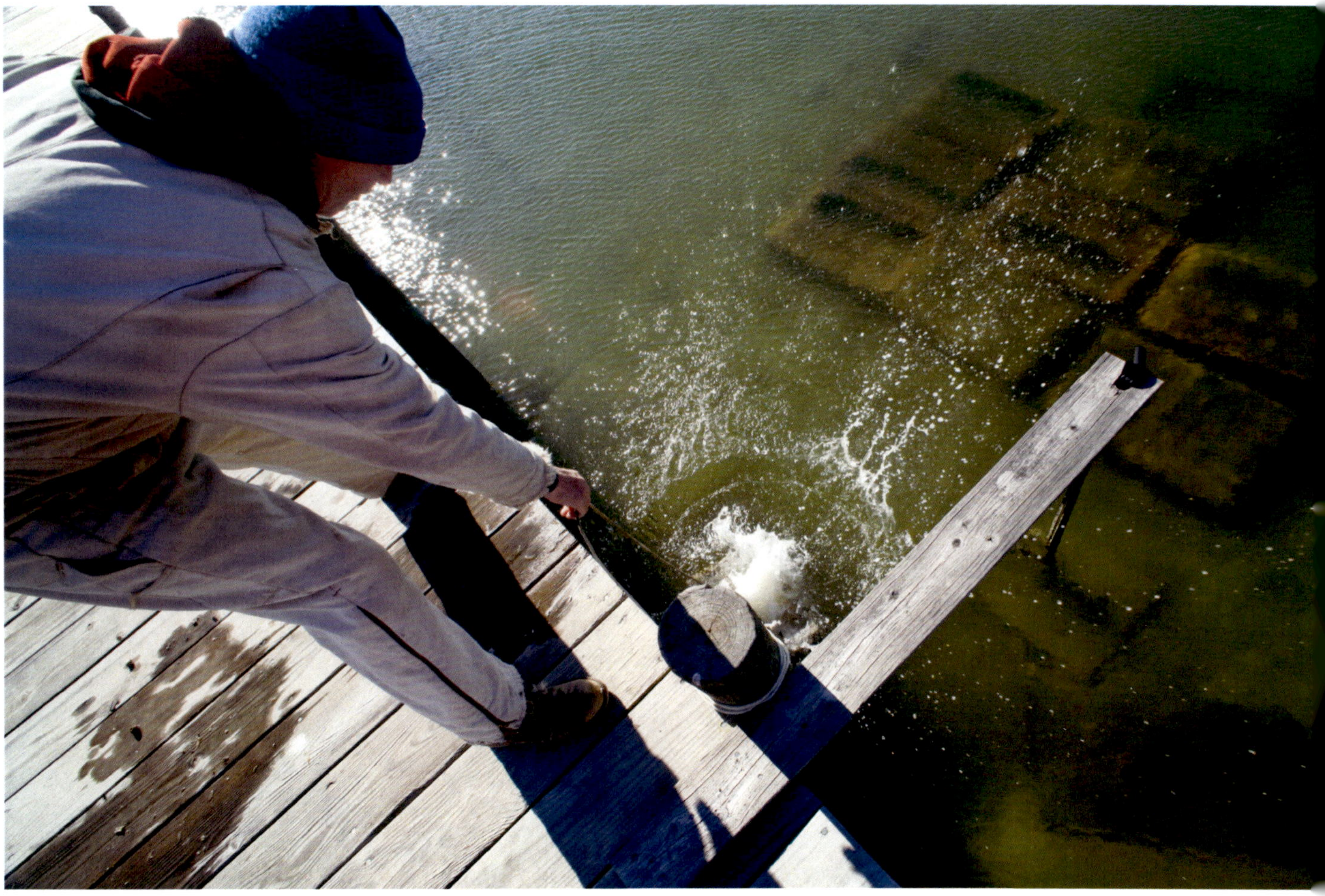

Oysters are a keystone species of the Chesapeake Bay and are disappearing quickly. They provide reef habitat and filter out pollution—helping to keep the bay healthy. One of the few ways to maintain the oysters in the bay is by raising them for locals to eat. Peter Perina has been doing this for over 10 years at Eastfield Farms, where he maintains a healthy oyster population to feed Virginians while also sustaining Chesapeake Bay habitat.

Victory Farms is getting ready for the market season, planting lettuce, and receiving mulch and compost from Hanover County to help fertilize rows and cover walkways. Hanover County collects organic material throughout the year and provides compost and mulch to several farms in the area.

Victory recently doubled its acreage and CSA membership, and despite this growth there is still a waitlist to join their CSA and receive produce—such as the newly planted lettuce to the right.

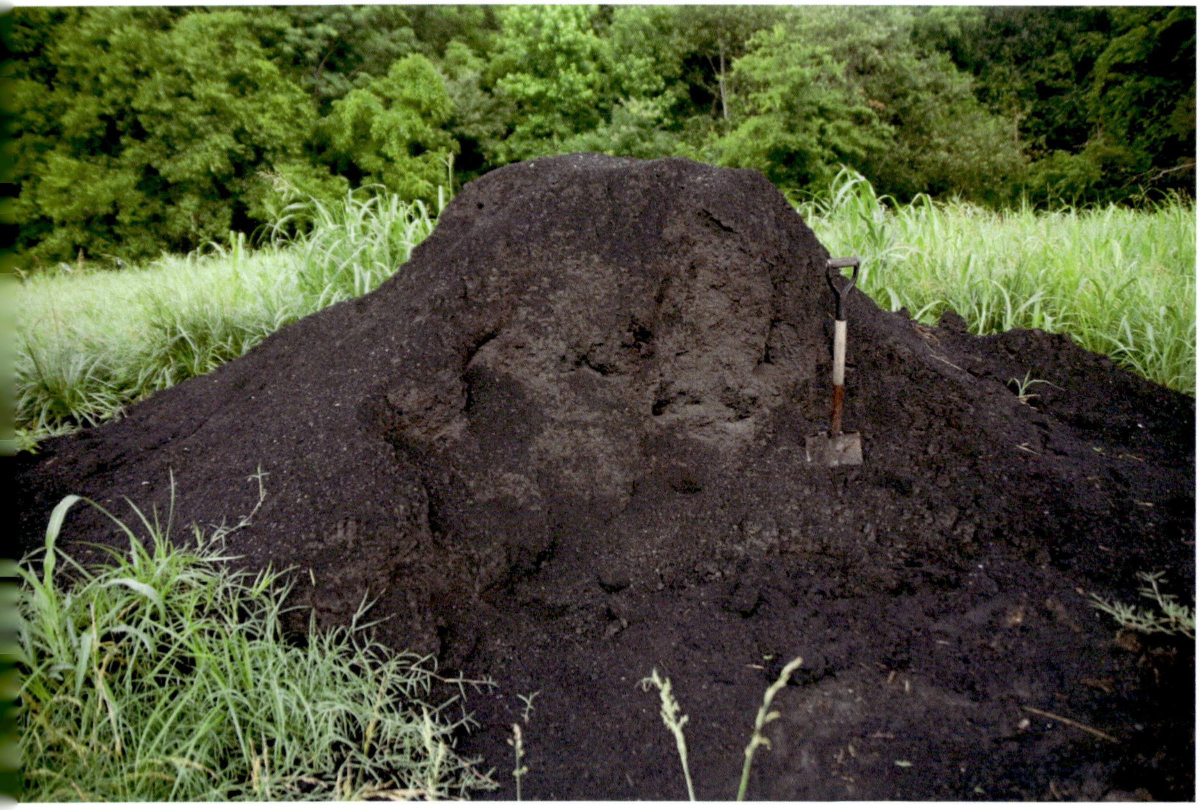

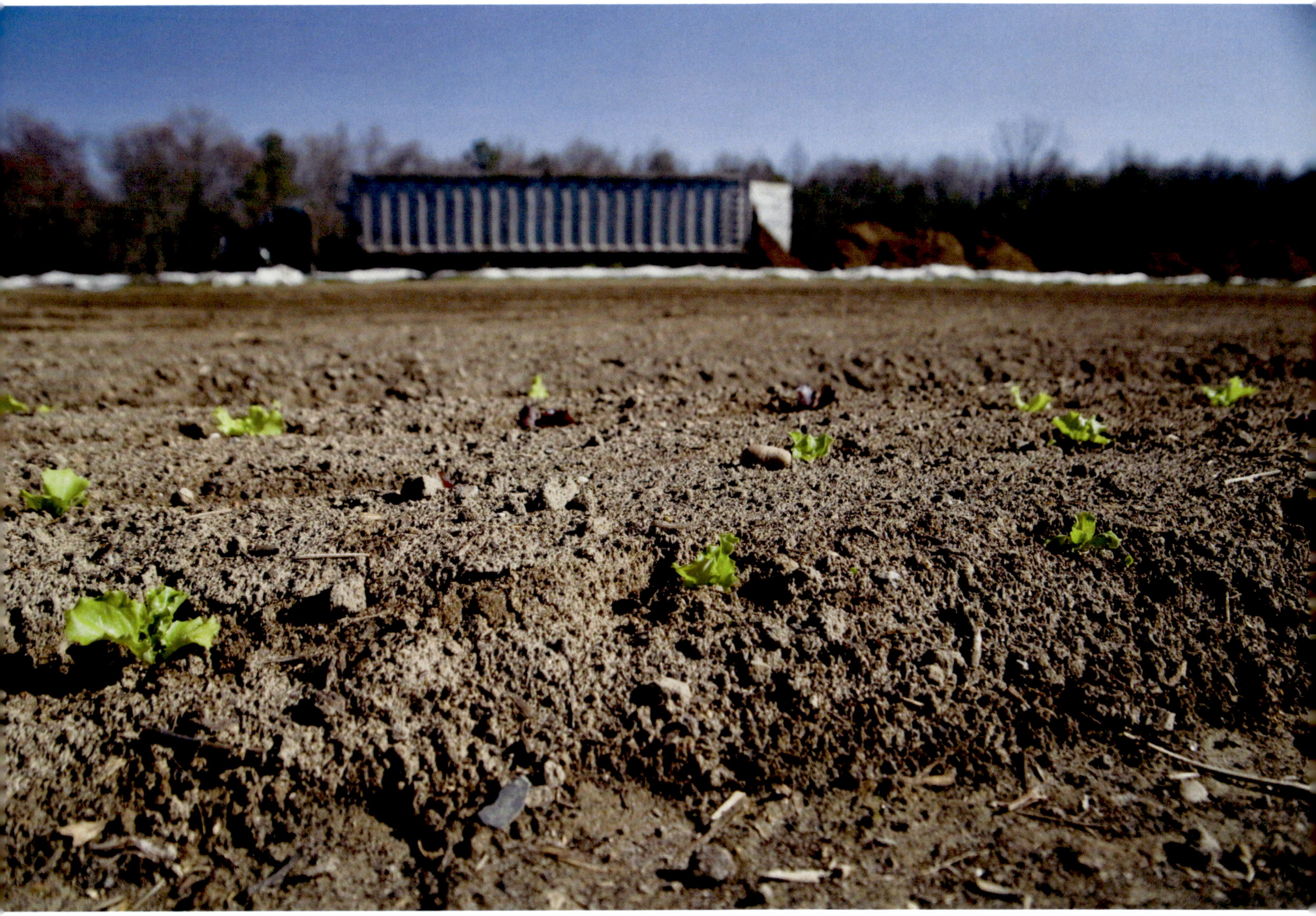

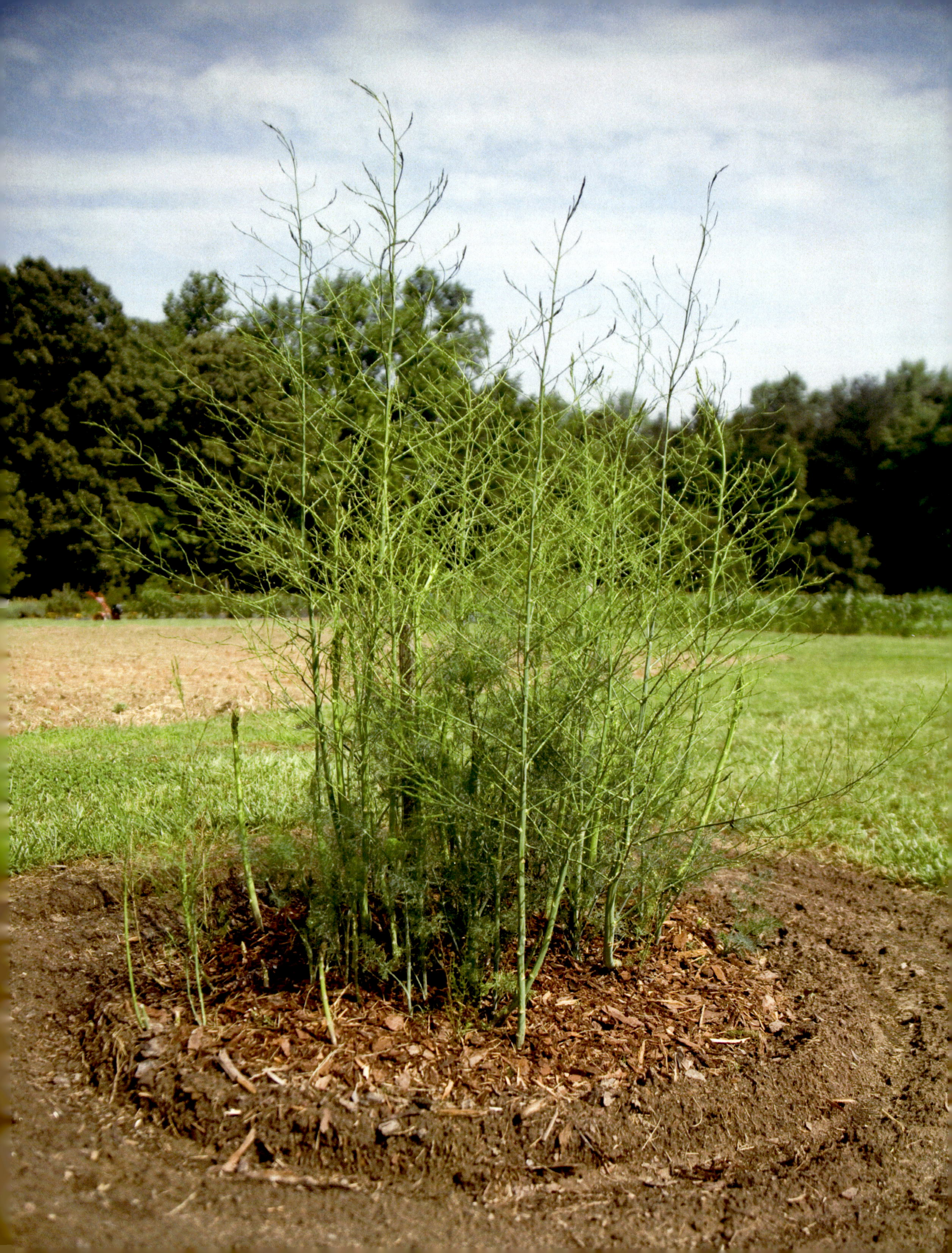

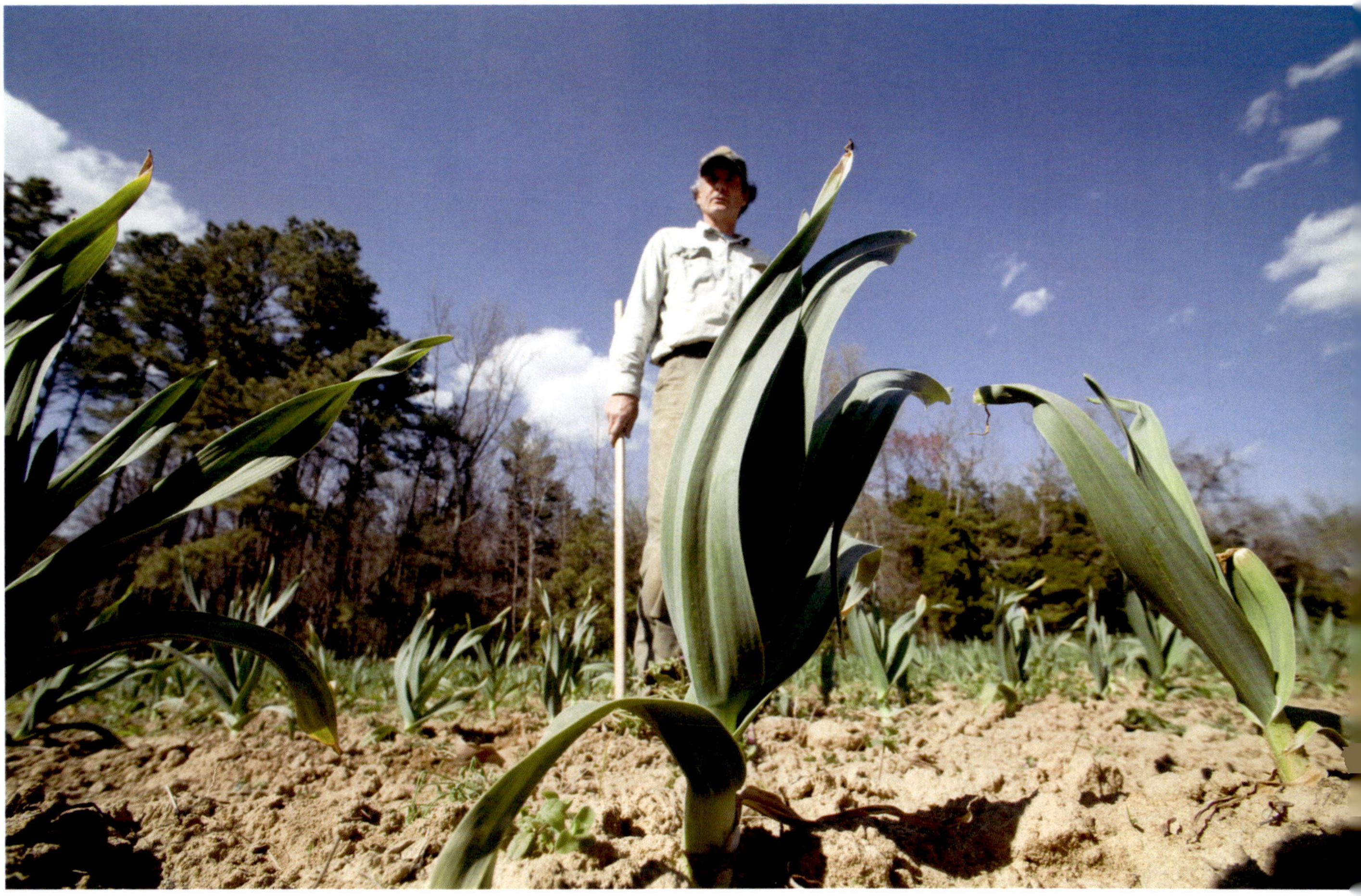

Asparagus is harvested in early spring and then left to grow for the rest of the season. This asparagus patch was not planted by farmers, but by birds sitting on a fencepost that used to be there.

Lawrence tills elephant garlic on his antique tractor. Blenheim is one of the few certified organic farms covered in *Groundwork*.
Very few growers in *Groundwork* are certified organic because of the expense and paperwork involved—and they sell a majority of their produce to local communities that can visit the farm and learn if a farm is using sustainable or organic practices. Blenheim, however, is dedicated to organic farming practices, and the certification is still a selling point.

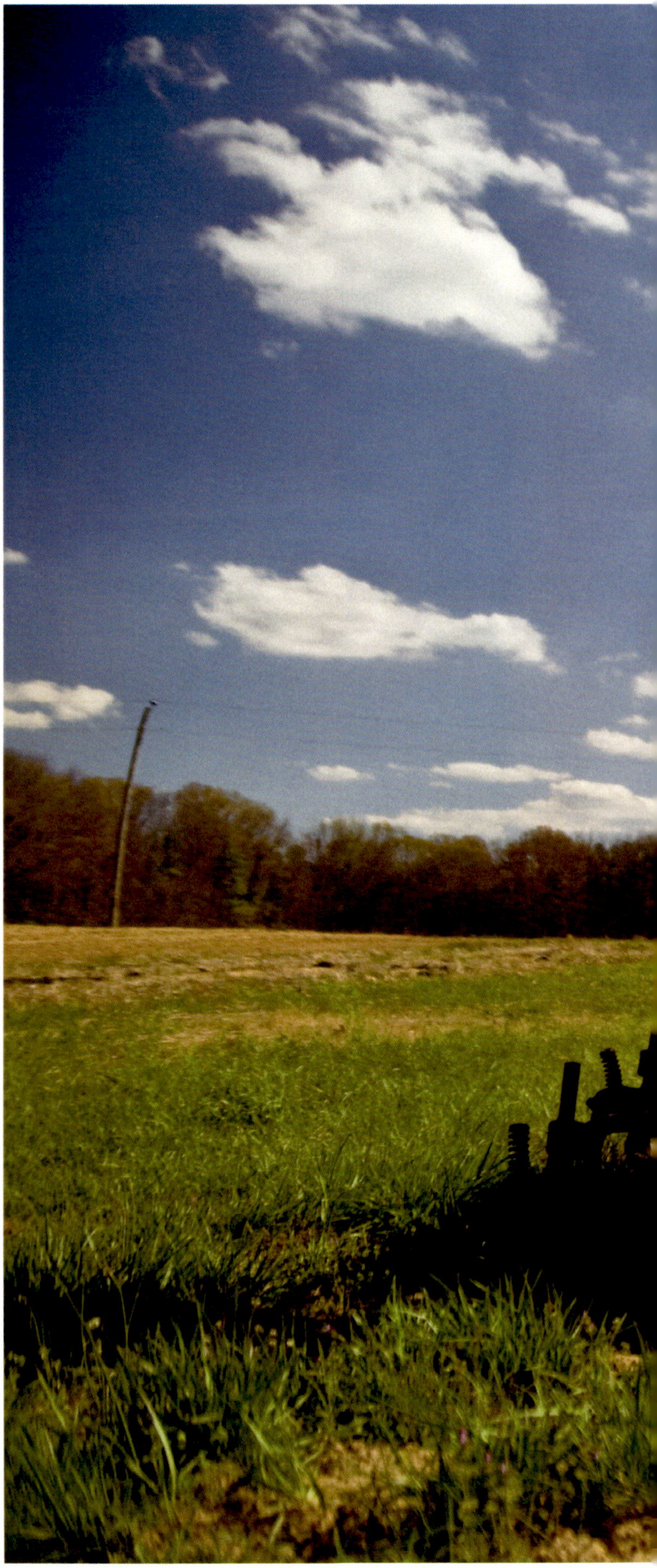

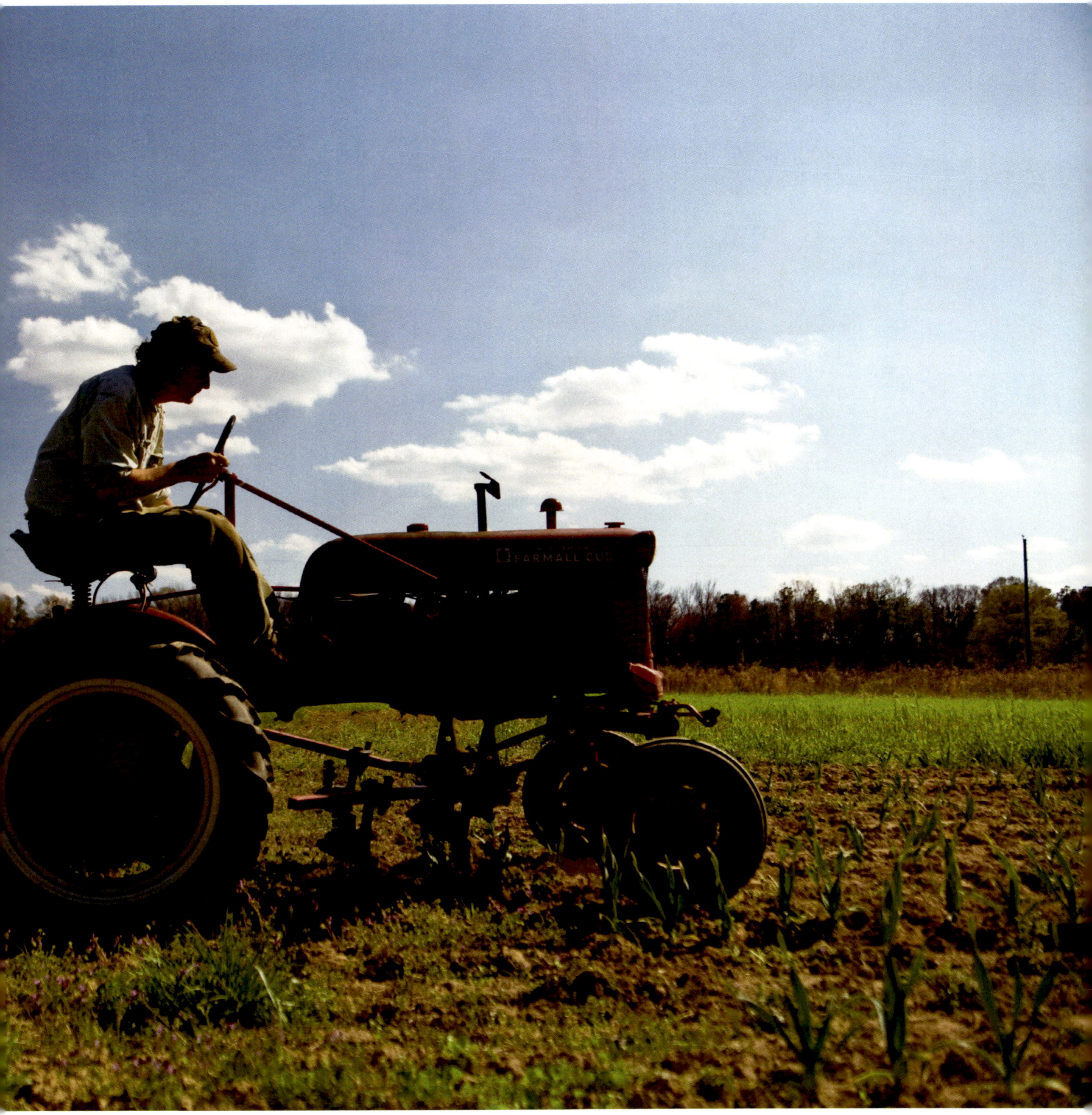
FARMALL CUB

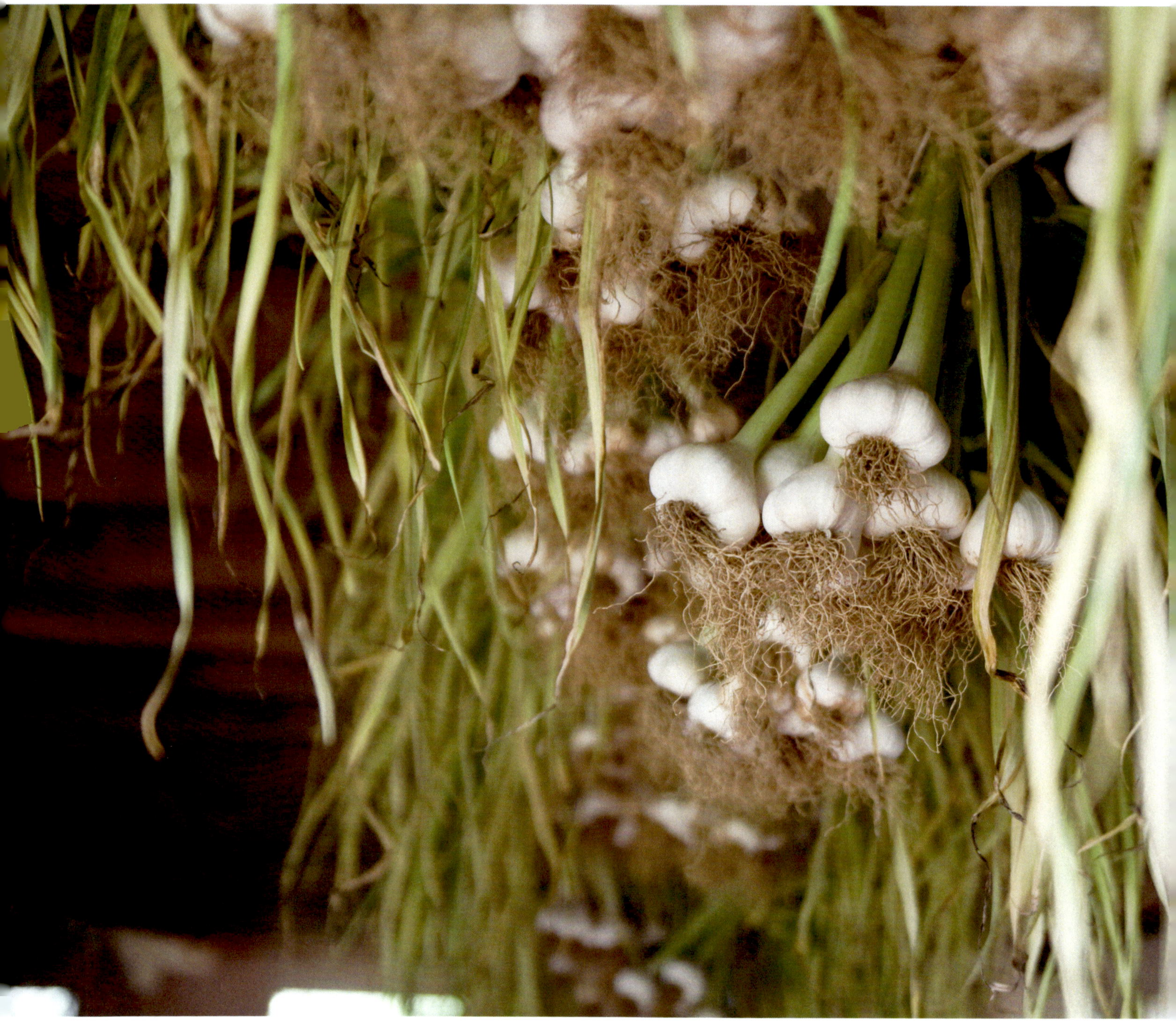

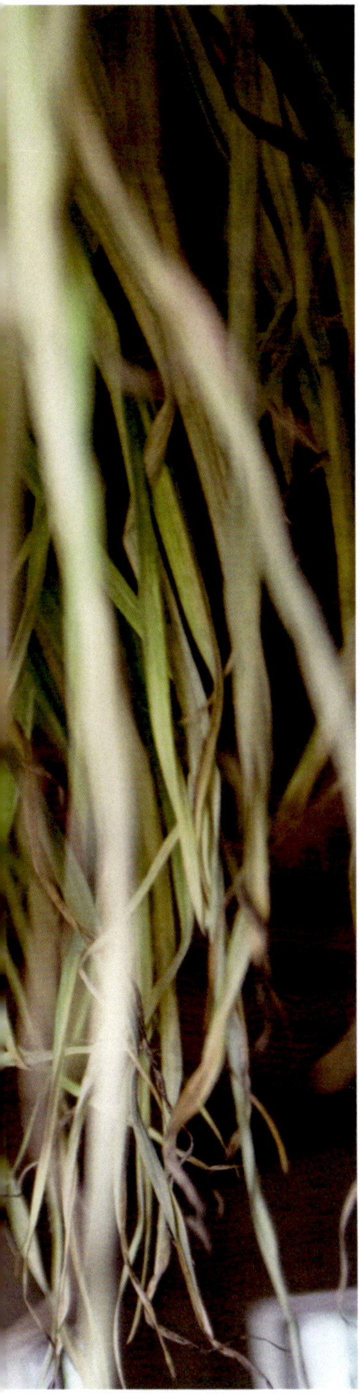

Garlic hangs in Fertile Crescent's barn. The garlic was hung after harvesting in the spring to dry out during the summer and then be sold at market. Many farms also sell freshly picked garlic in the spring, but this garlic is milder and does not have fully formed cloves.

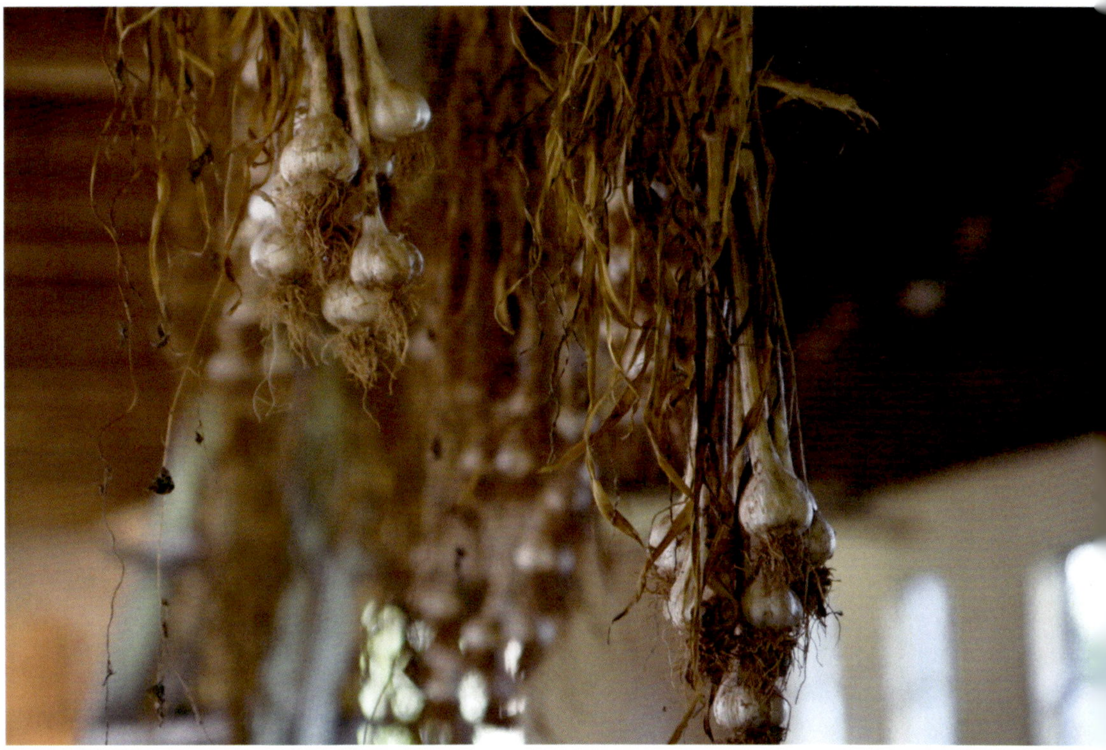

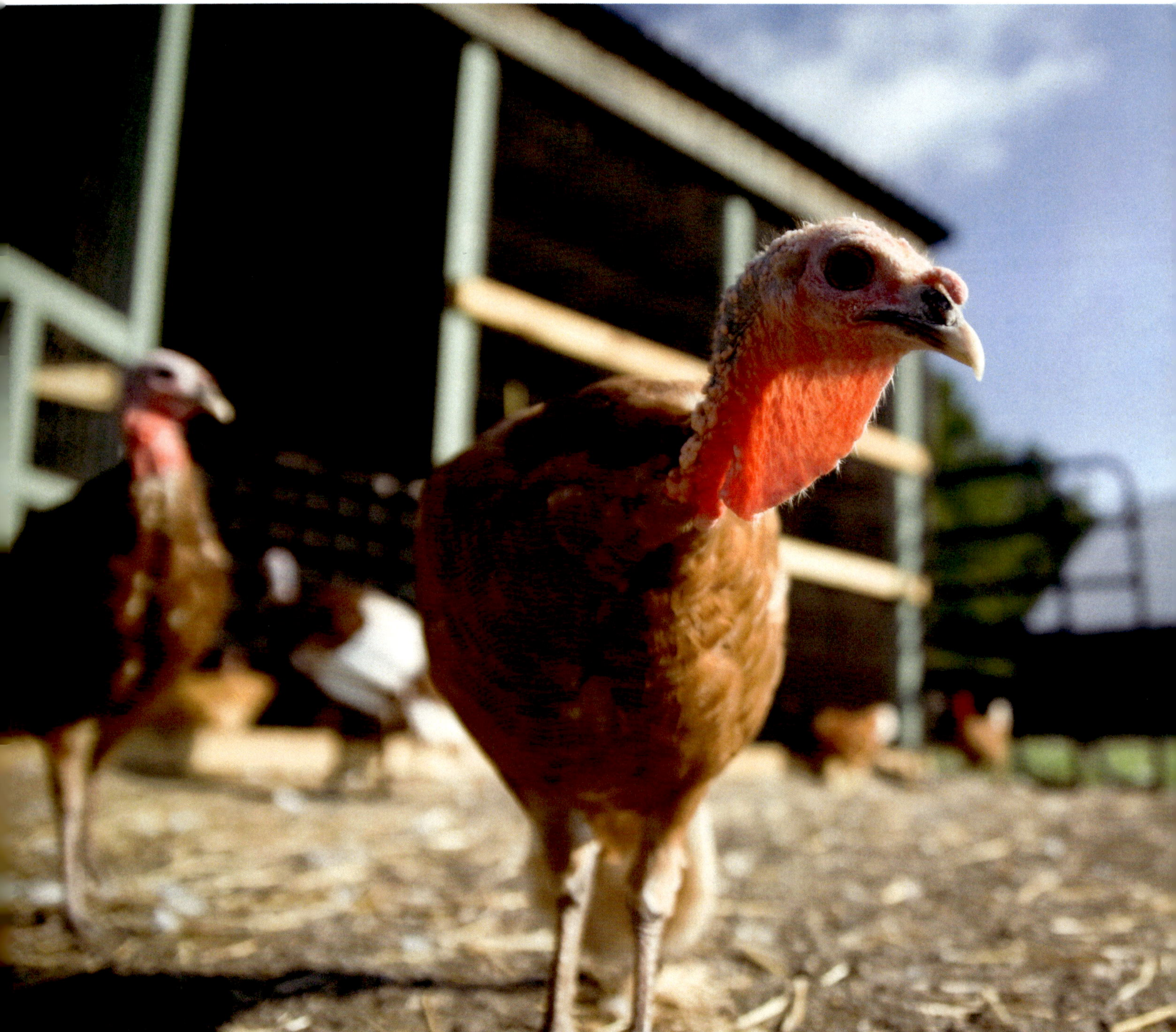

This turkey is being raised for Thanksgiving at Simply Abundant Farm, just outside of Richmond. Simply Abundant keeps the chickens and turkeys together in a mobile hen house amidst a flock of sheep. The turkeys at Simply Abundant do take longer to mature, but taste better, forage for much of their food, and receive individual care for any sickness.

Many farms keep beehives to help pollinate their plants and provide a little honey. They have to keep the bees busy starting in spring. This bee is visiting some arugula flowers on Blenheim left over from last season. Many other farms also plant flowers to keep pollinators busy.

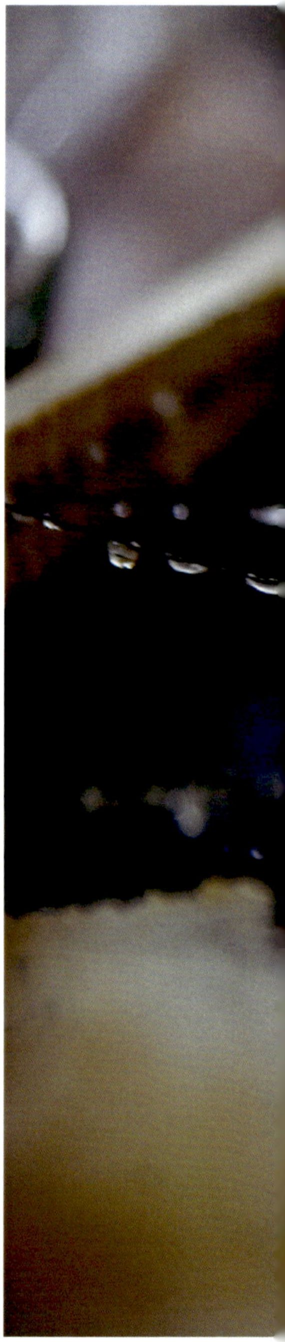

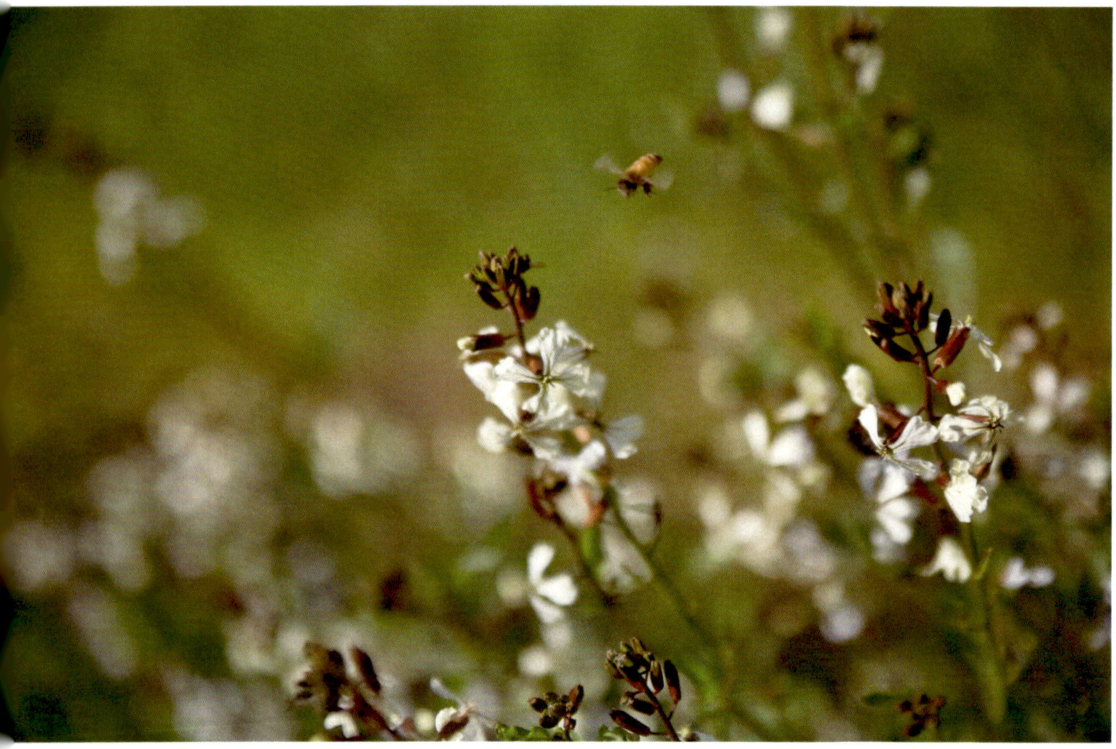

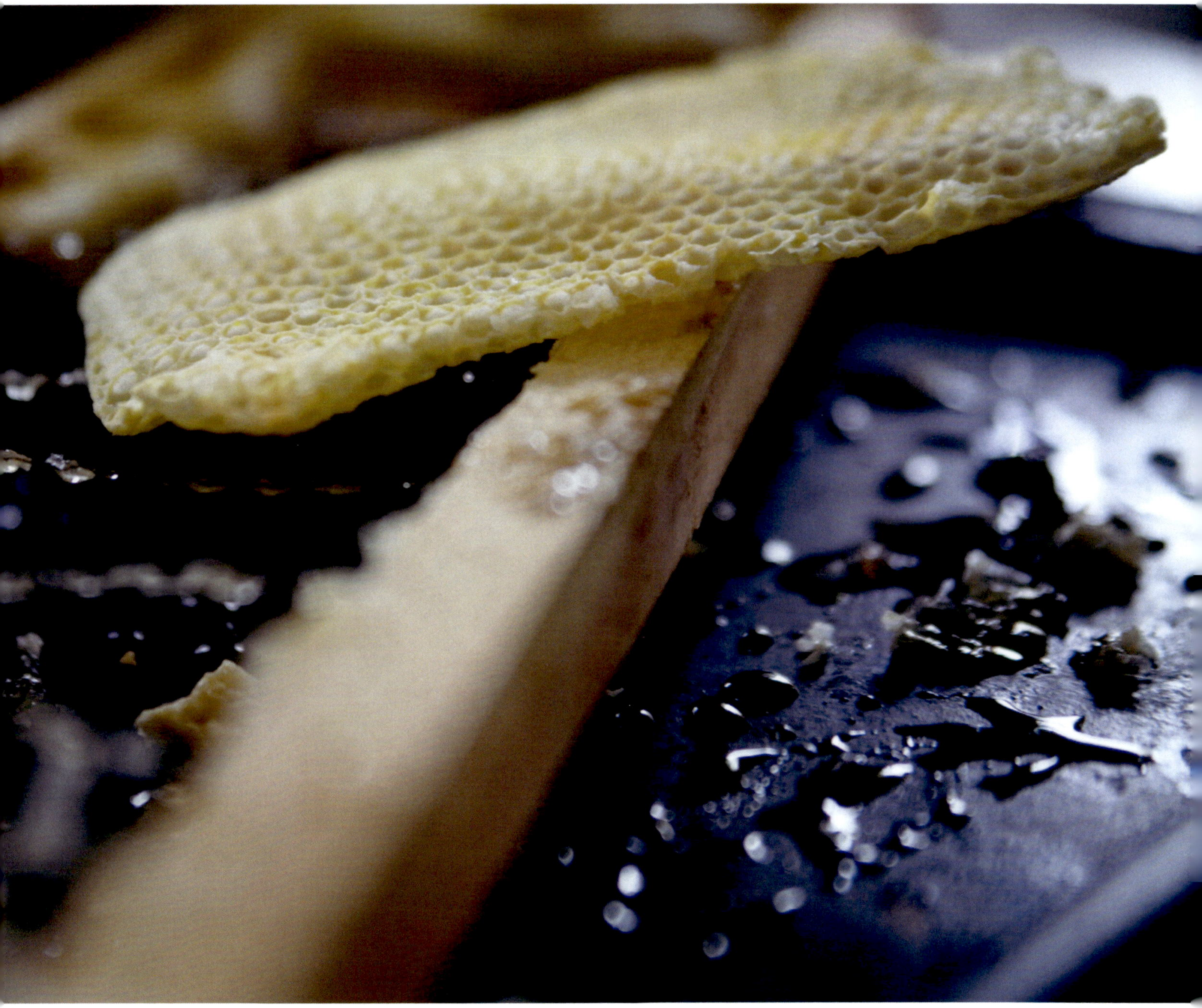

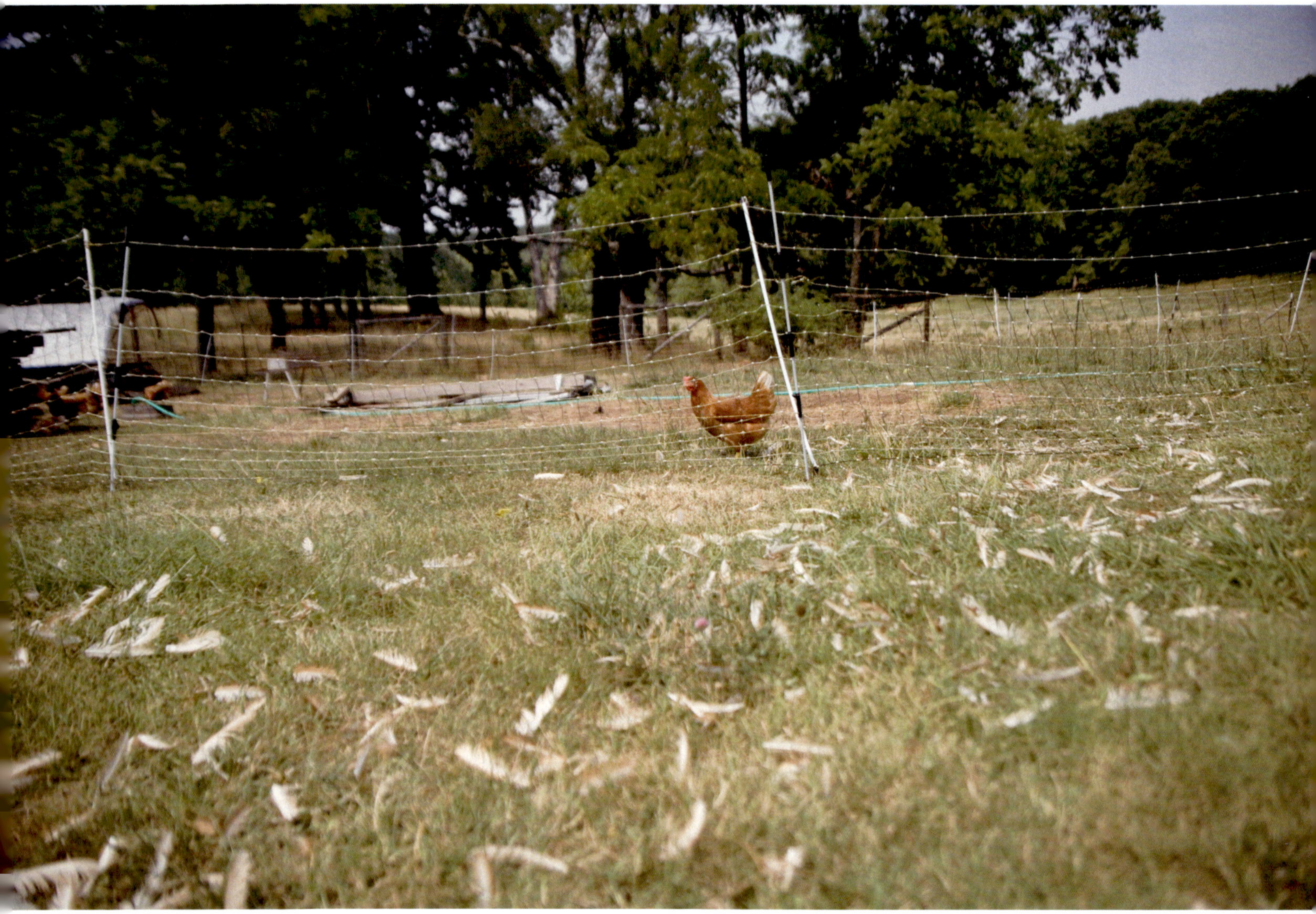

Clipping chickens' wings keeps them in their pen, away from predators and near their laying boxes. This chicken was not so lucky when it escaped from the chicken coop—it was caught by a predator before the guard dog jumped into action.

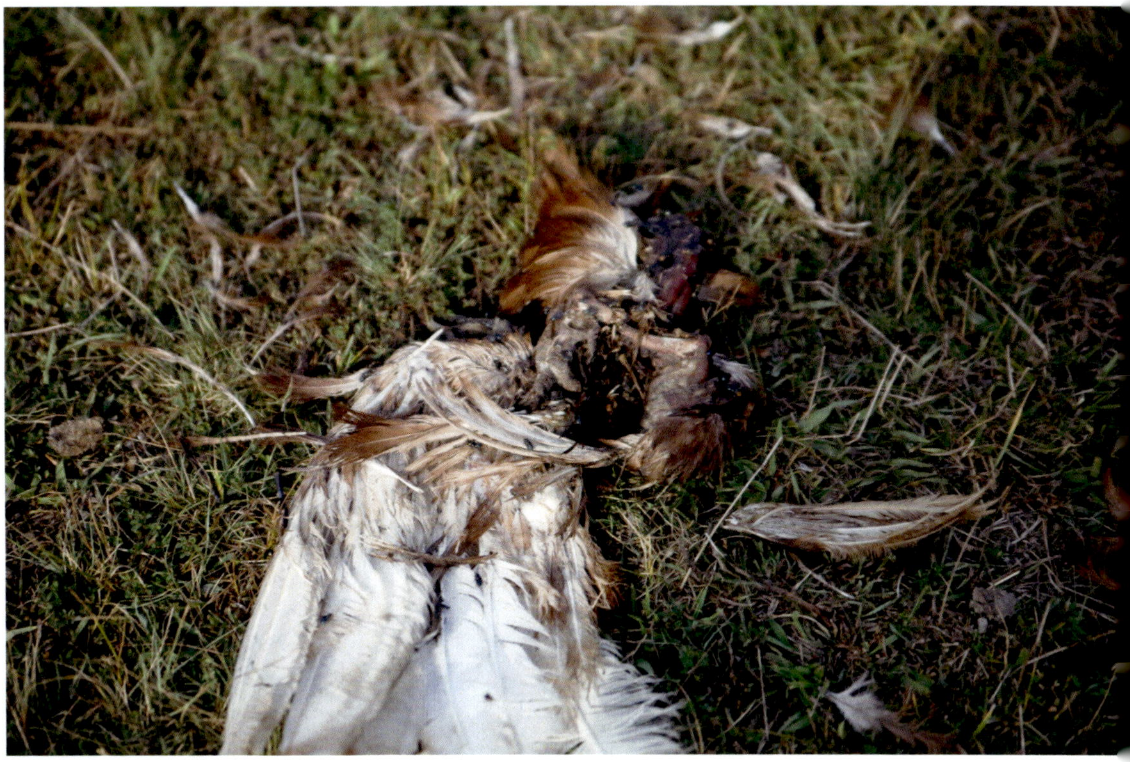

Faith Farm has plenty of pasture for their livestock. They rotate their livestock to preserve each section of pasture and let it recover so that the land is not overgrazed. The barn provides shelter and shade for sheep and cattle in the summer.

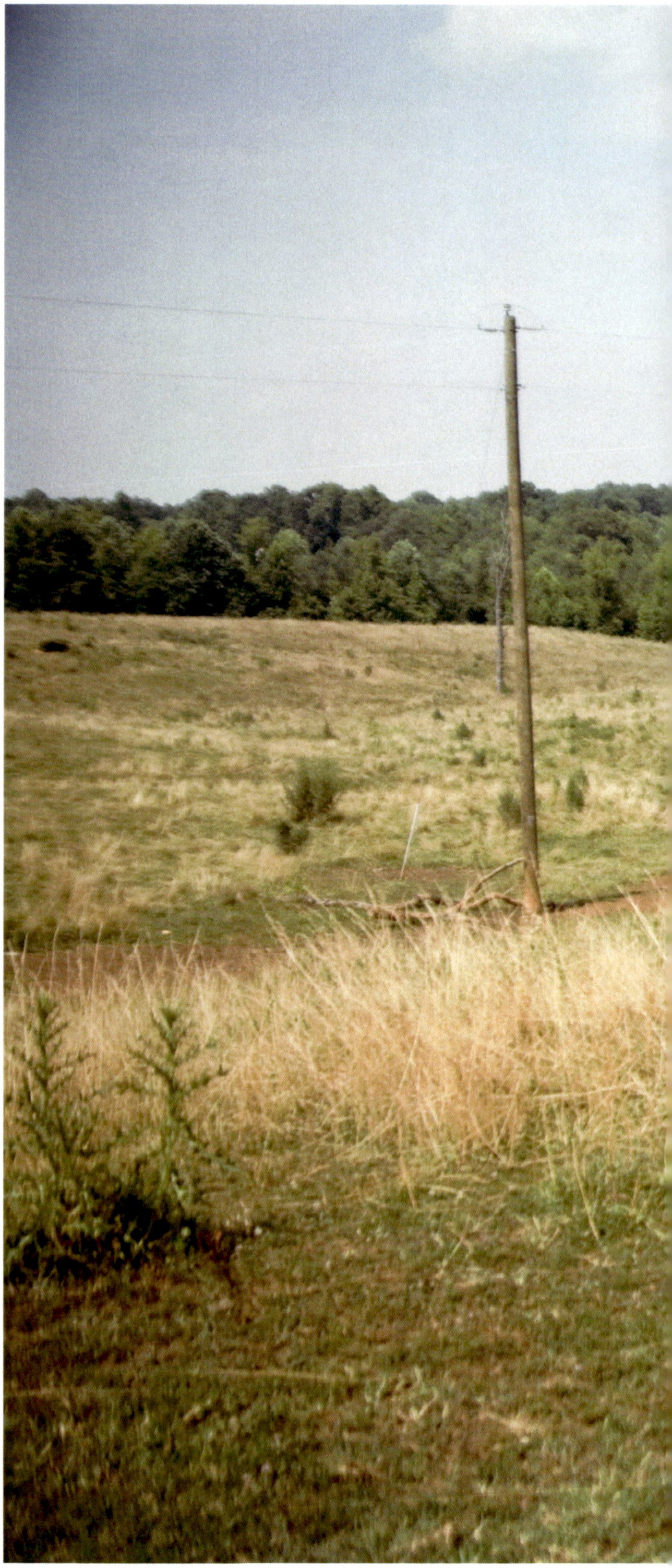

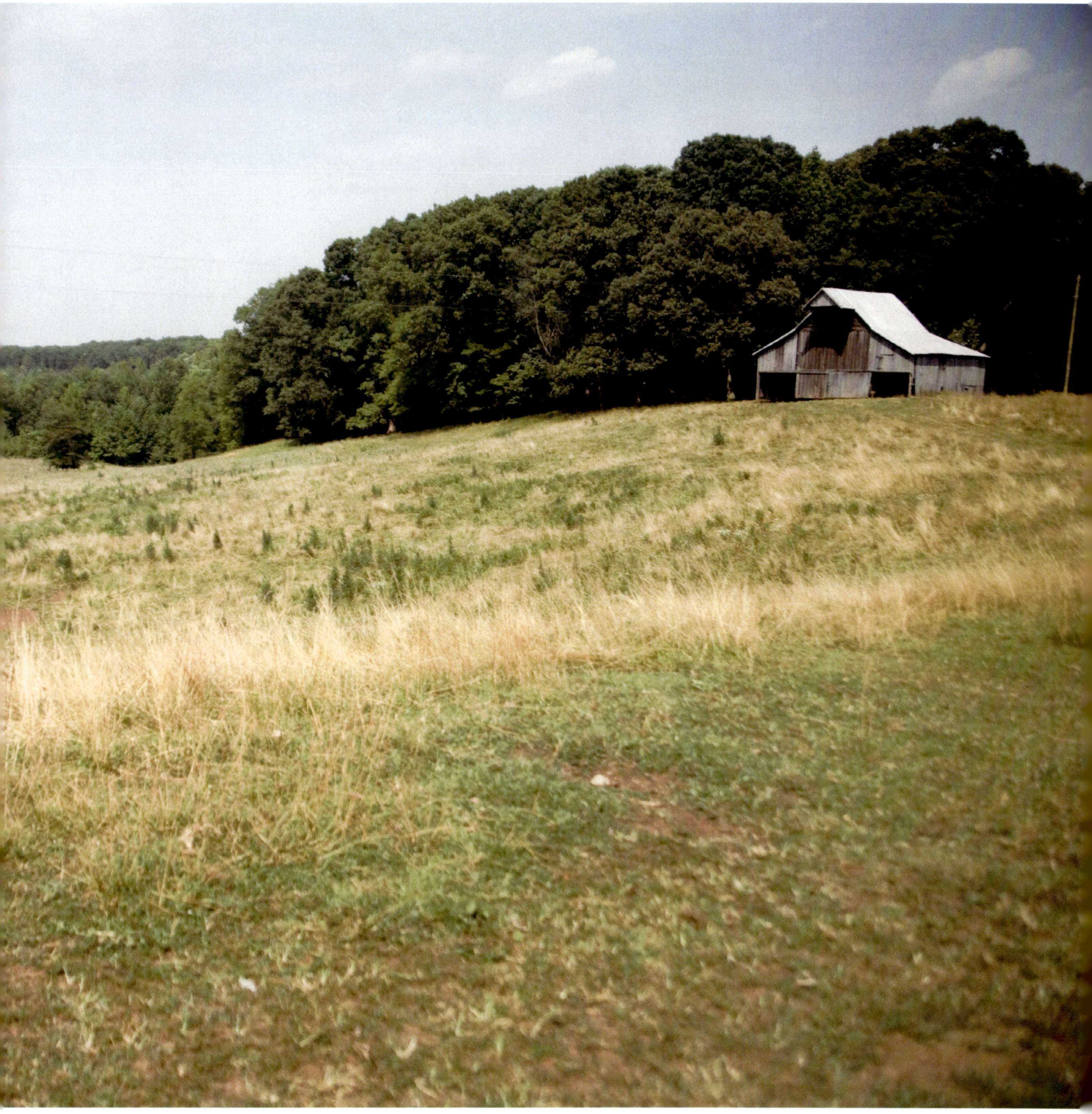

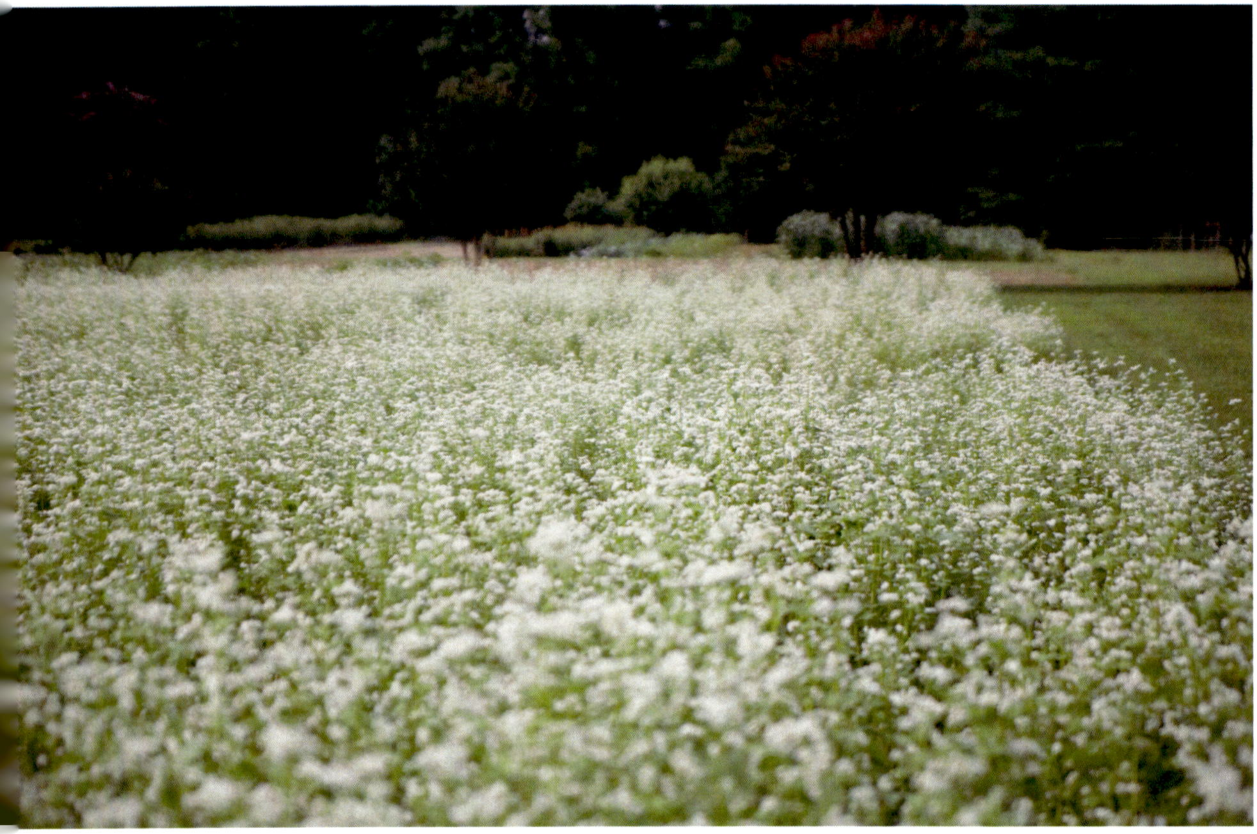

Red Dog Farm is a small operation of a lifelong grower—one who used to eat only food he grew. There are two main growing fields and one is always under cover crops to rejuvenate soil nutrients. The soil is sandy and requires extra care to maintain nutrient levels with the least amount of fertilizer possible.

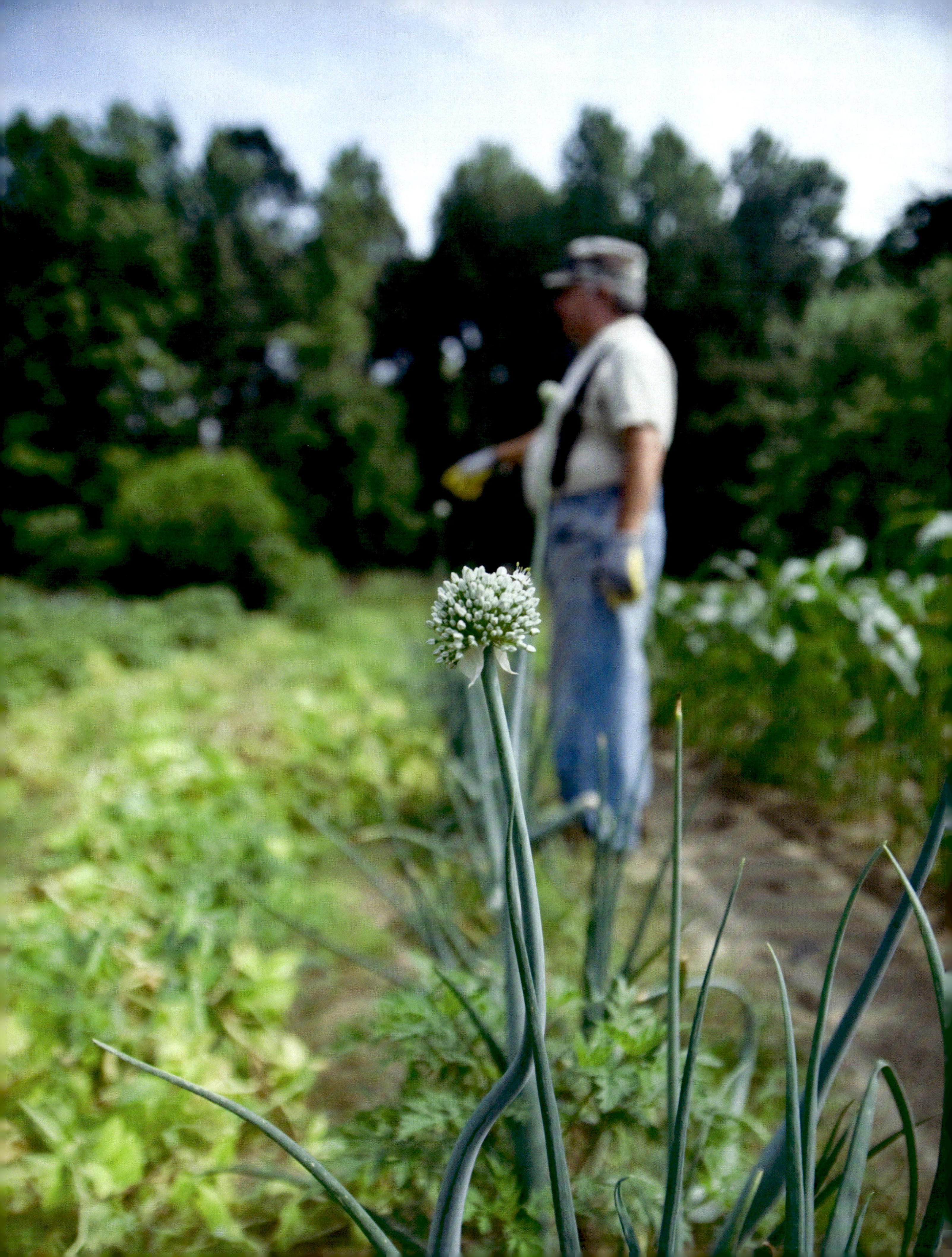

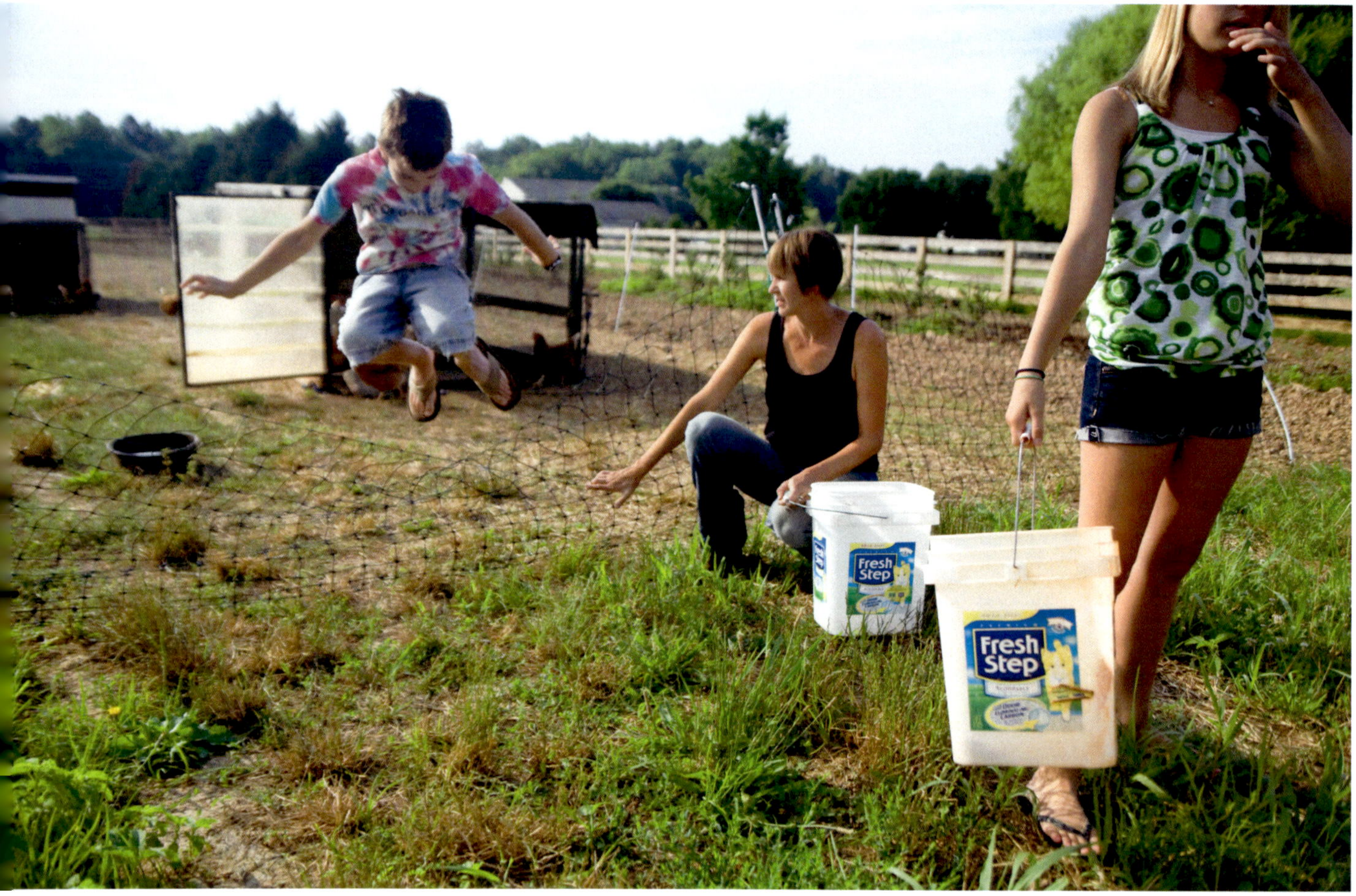

Trail's End Farm is a family farm just north of Richmond that is certified naturally grown. Certified naturally grown is a standard similar to organic certification that is policed by certified members—mostly different farmers—but requires less paperwork and expense. The certification is geared toward small farms that do not fit the model of farm for which the organic certification was developed (i.e. large organic farms that deliver produce nationally).

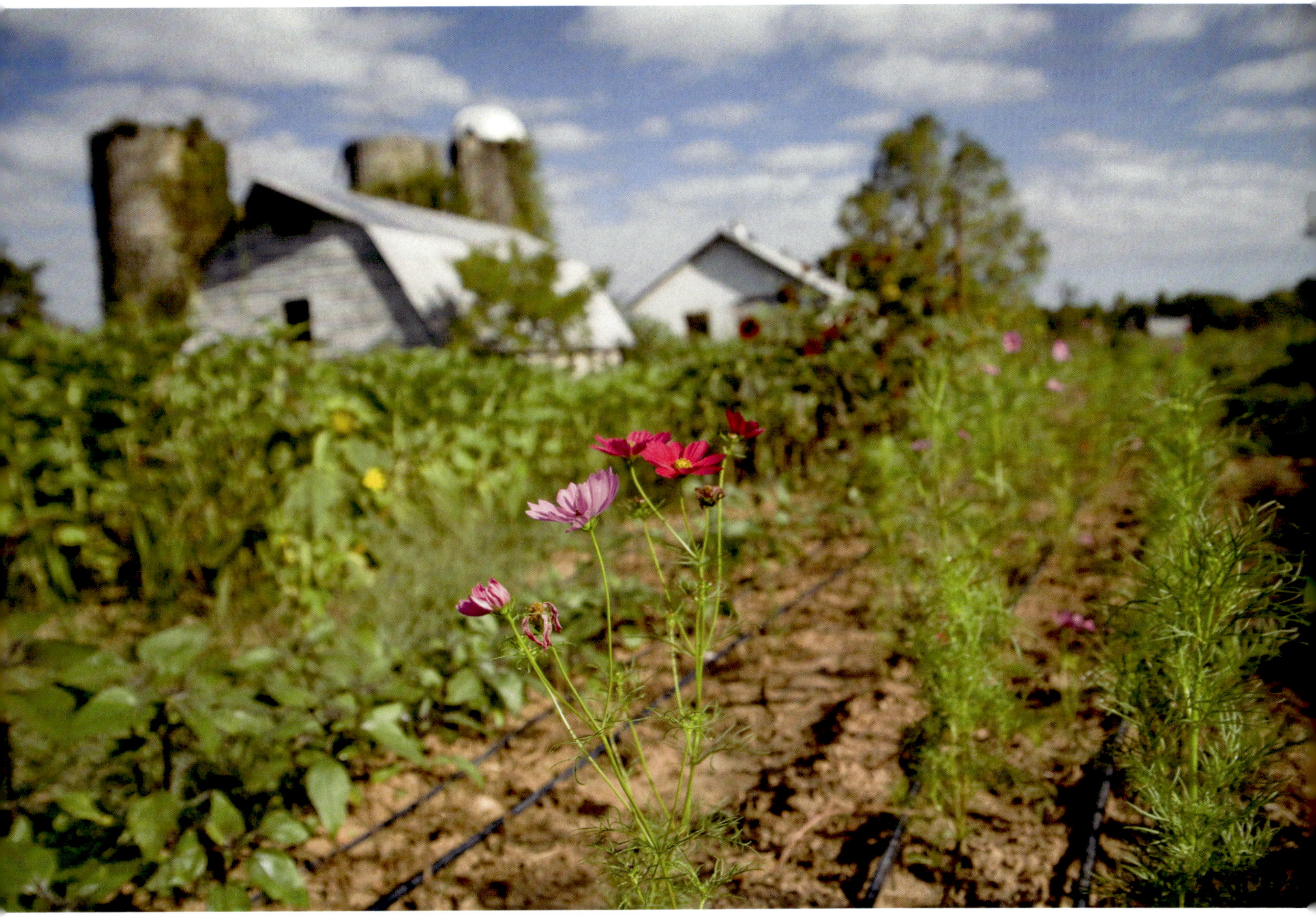

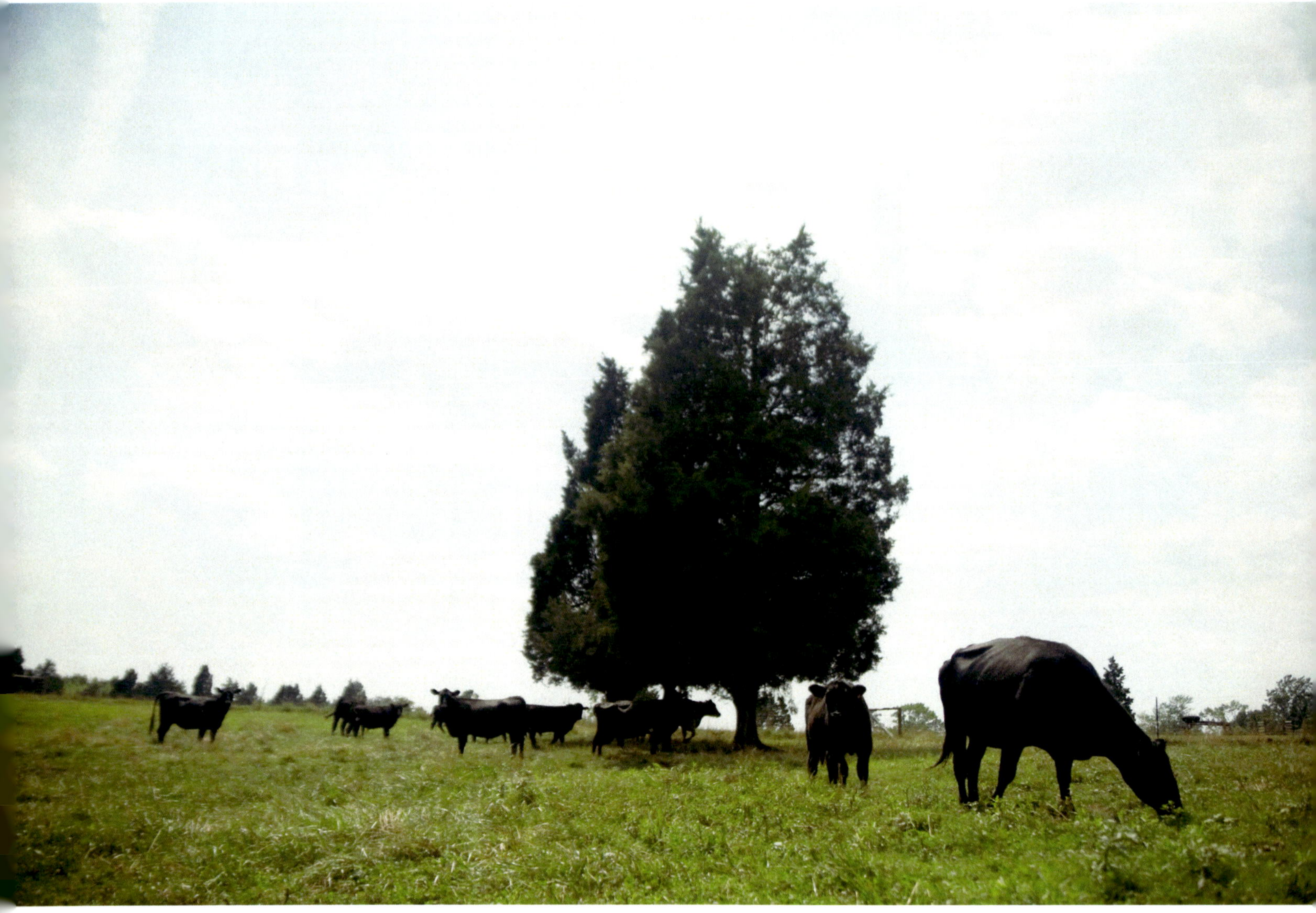

Above, cows graze in one of the fields at Faith Farm.

To the right, Faith Farm lets their smaller pigs roam freely, while the mothers and larger pigs are kept in pens—mostly to protect the farmers from a defensive mother, but also to ensure they don't lose a pig that is almost ready for market.

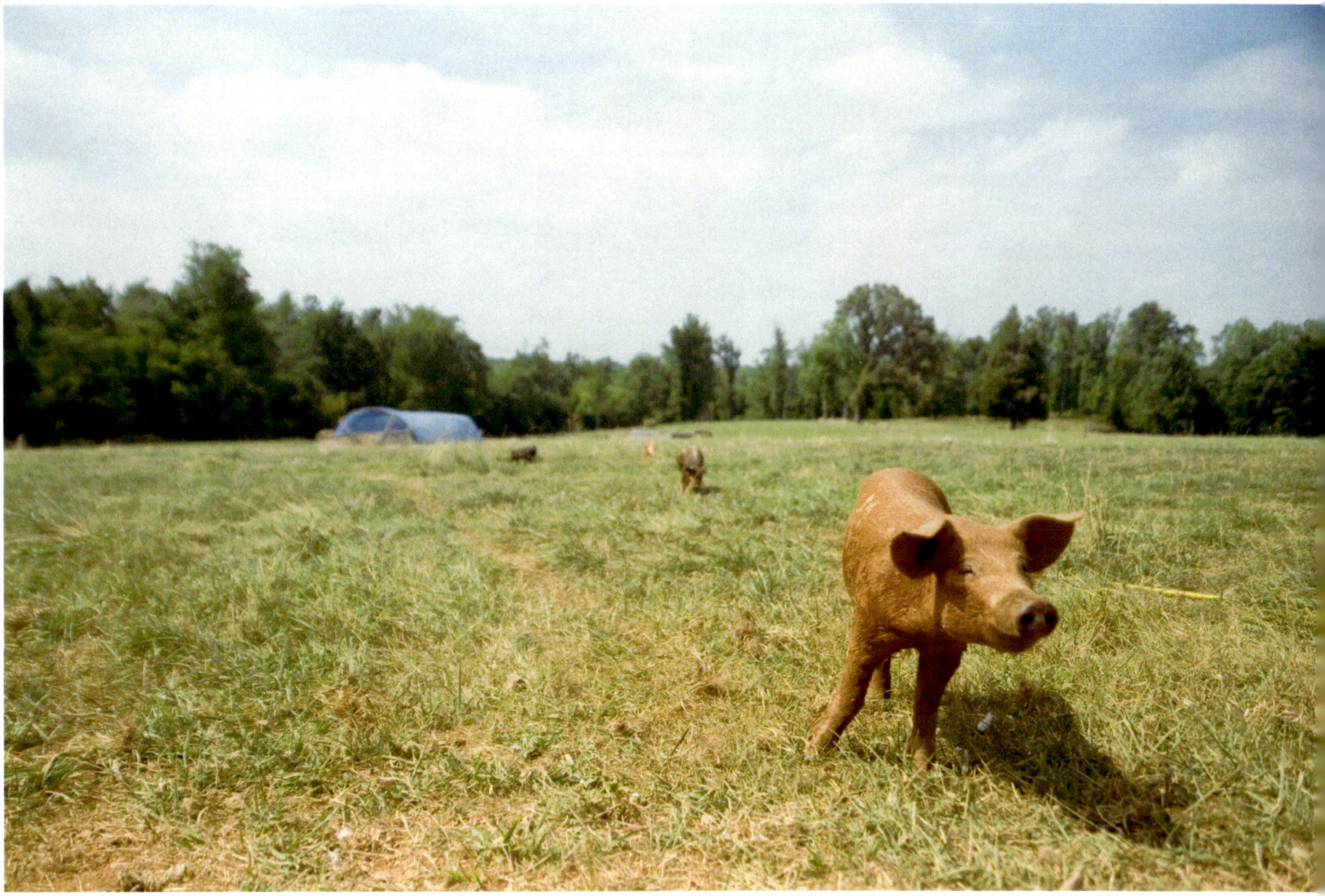

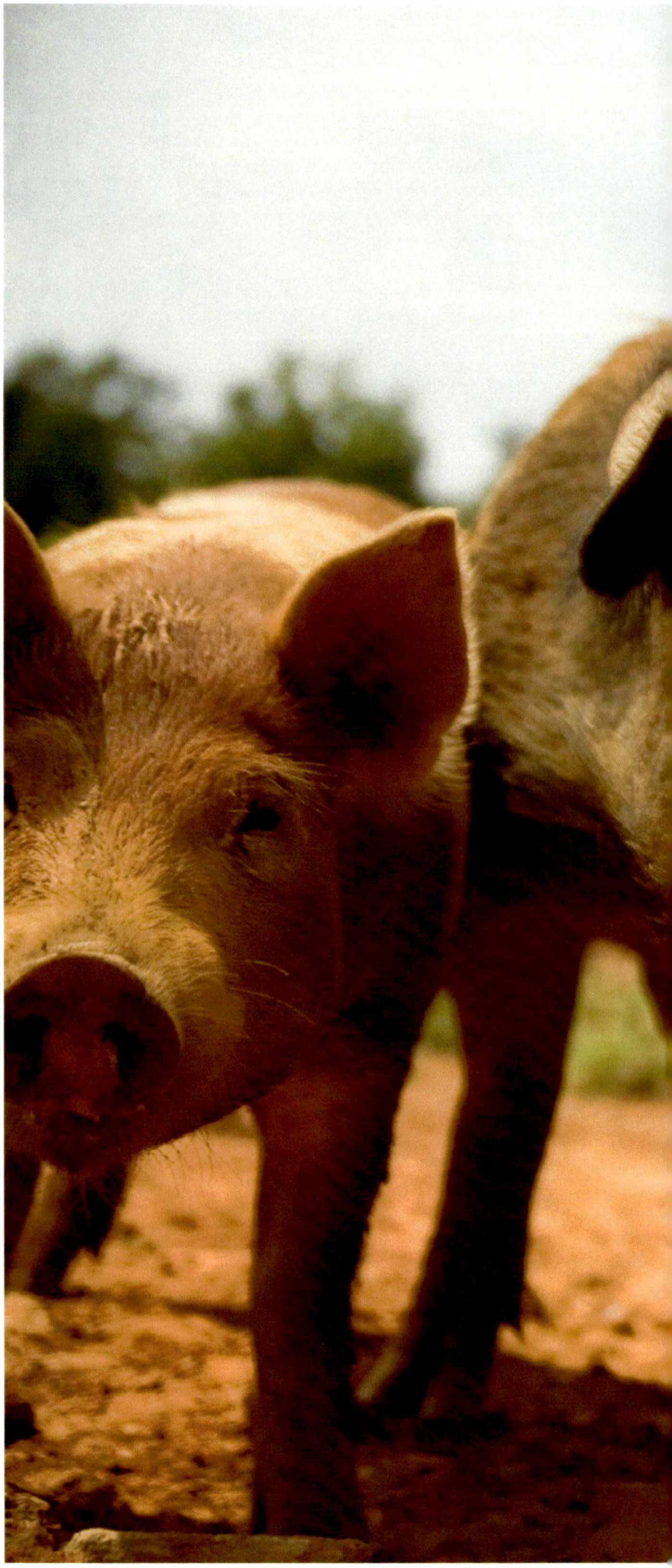

Curious pigs in their pen at Faith Farm.

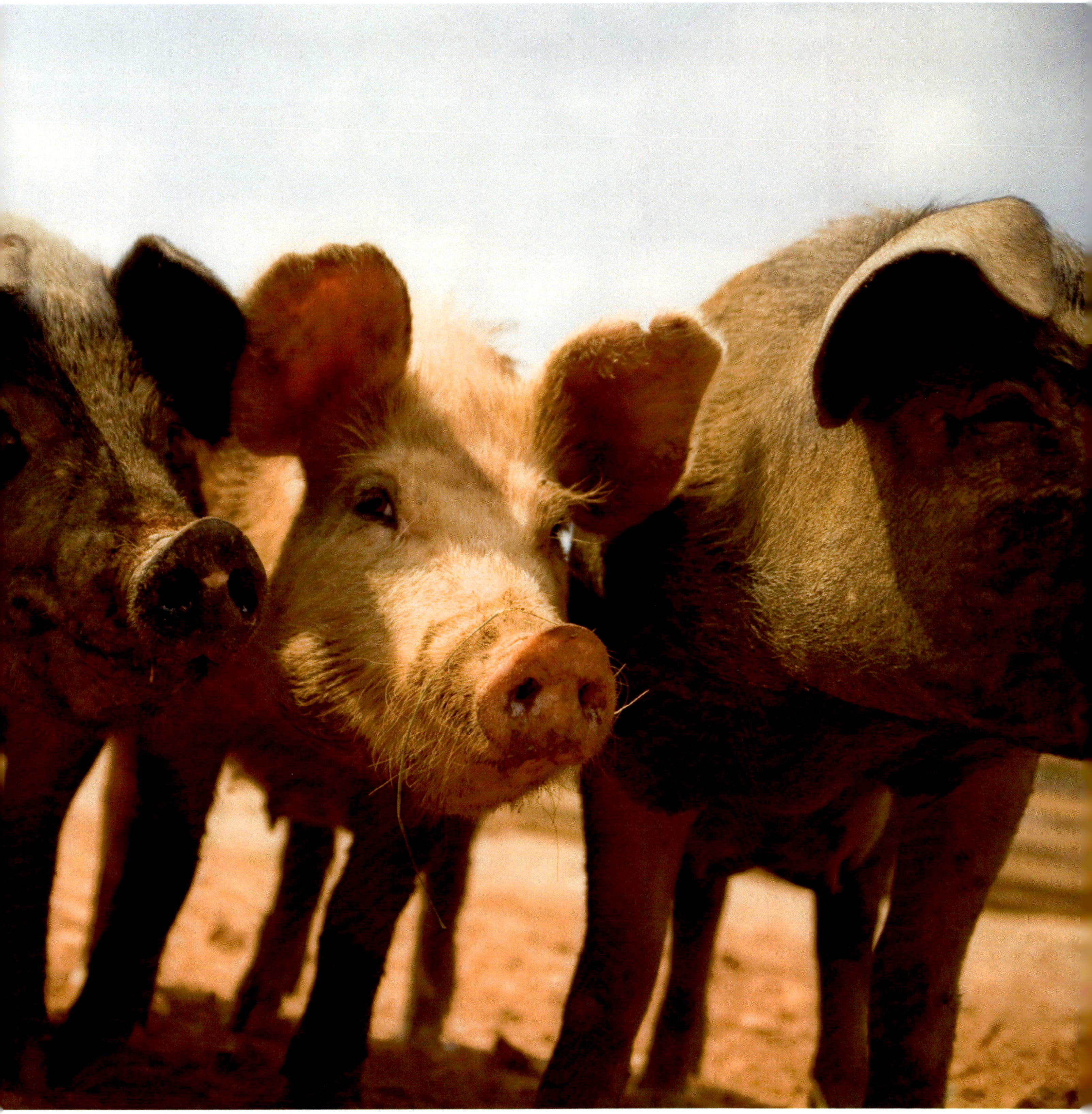

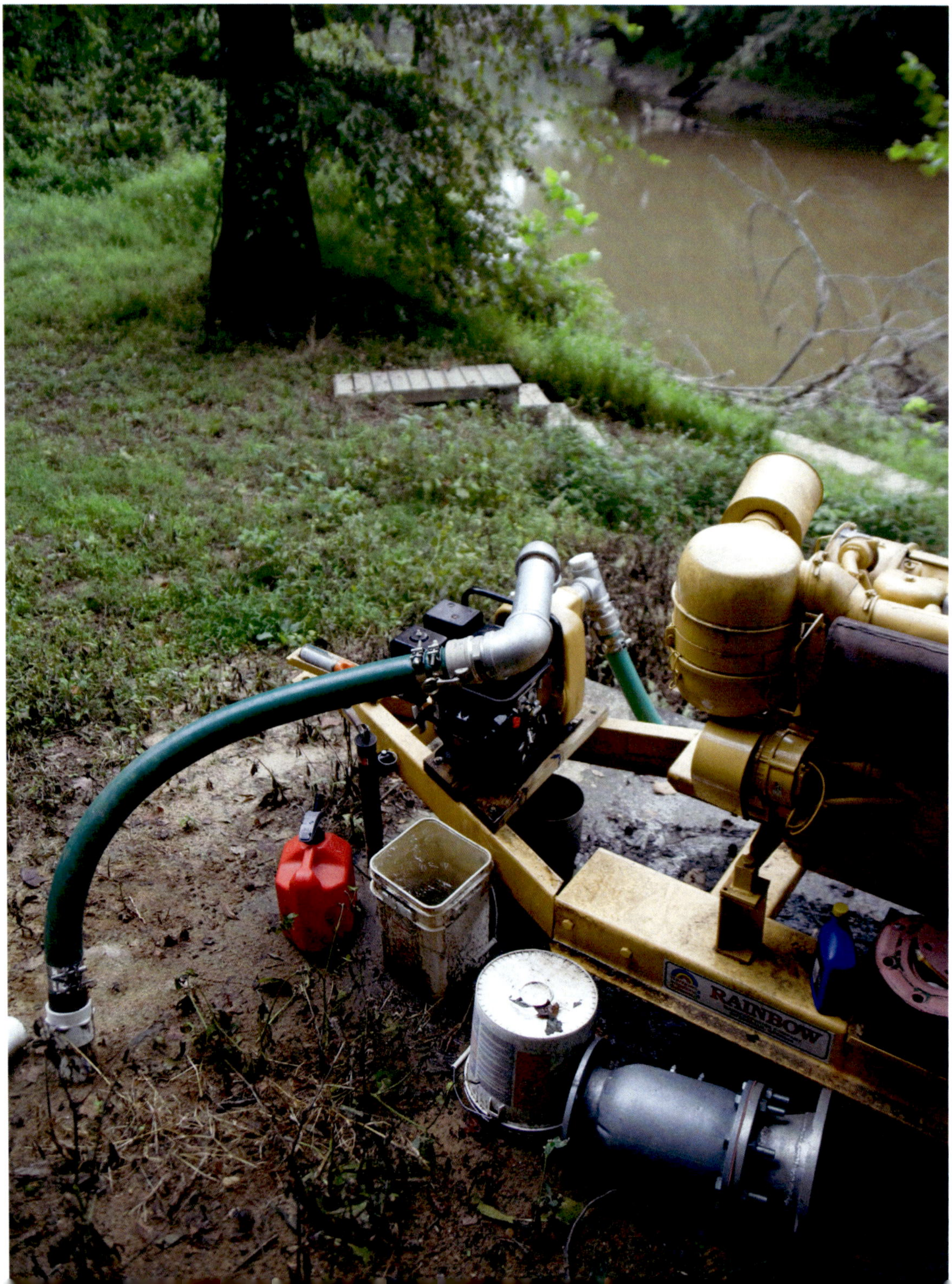
RAINBOW

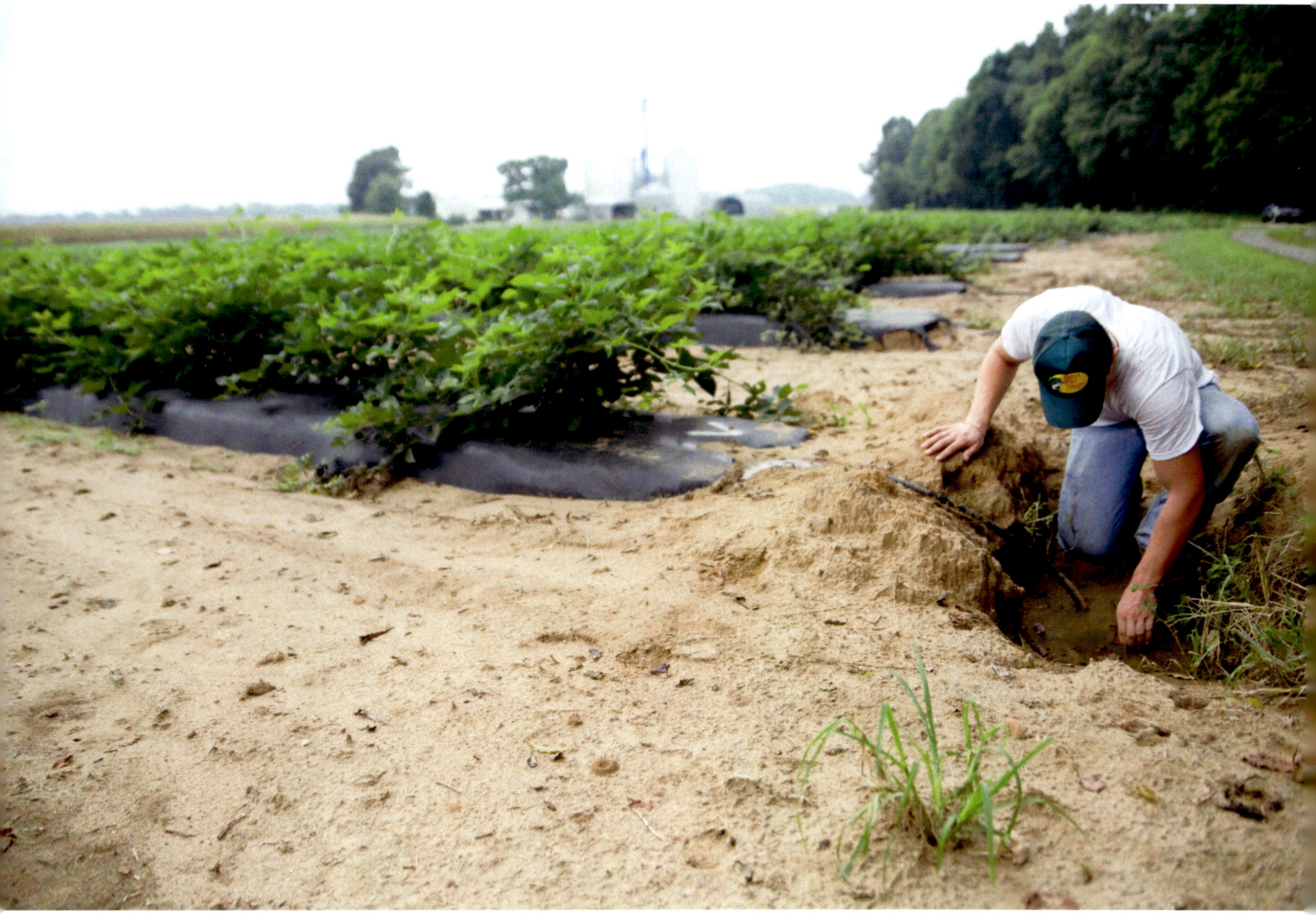

Agriberry draws water from the Pamunkey River and uses drip irrigation to water its berry plants. This allows the farm to use the minimum amount of water necessary, though it is more time-consuming because each irrigation hose must be checked to make sure it is working properly.

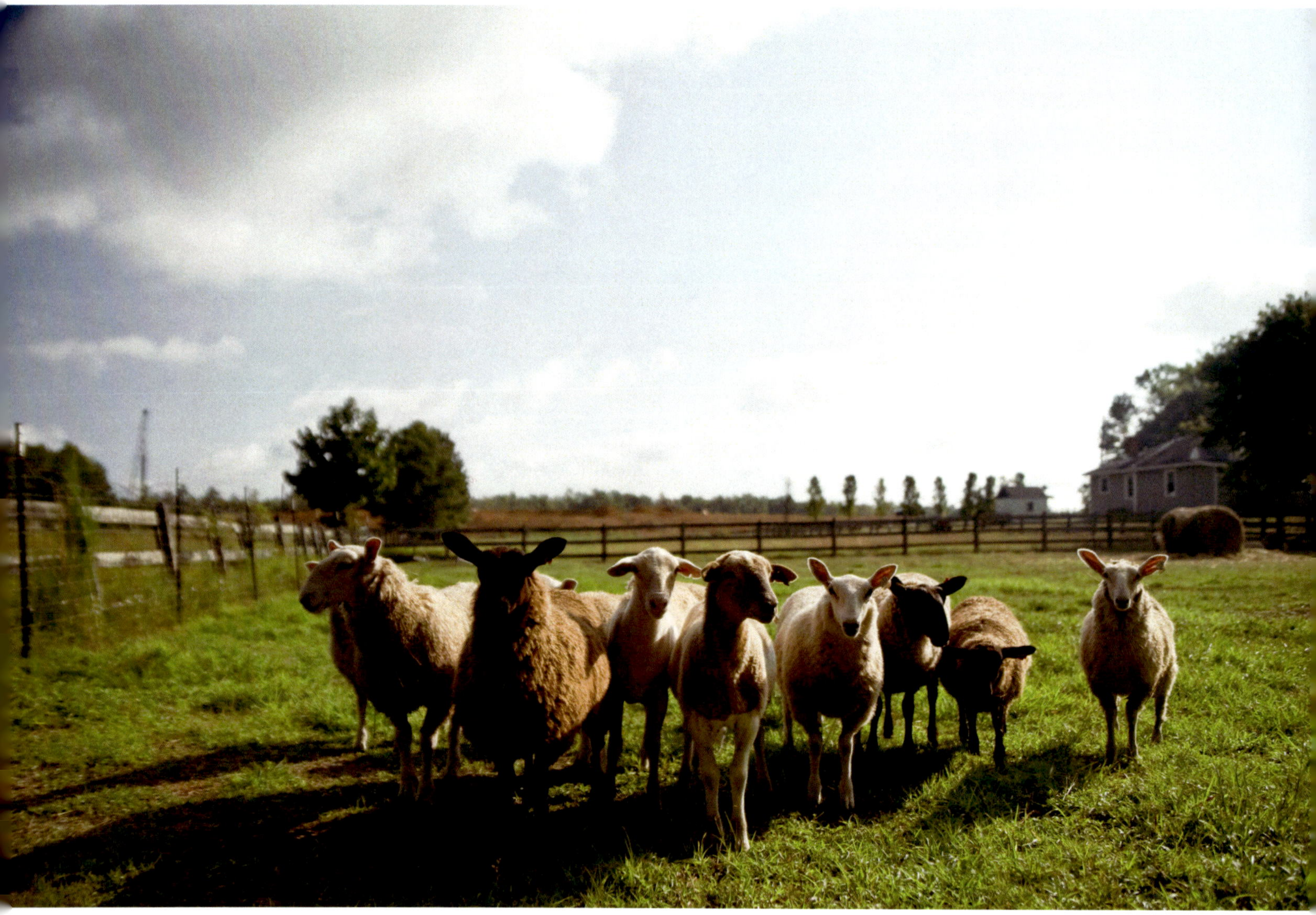

Simply Abundant Farm keeps three types of sheep and raises them for their wool and meat. Emily personally takes care of each sheep, tending to specific needs like injuries or illnesses while not subjecting the other animals to unnecessary care or antibiotics.

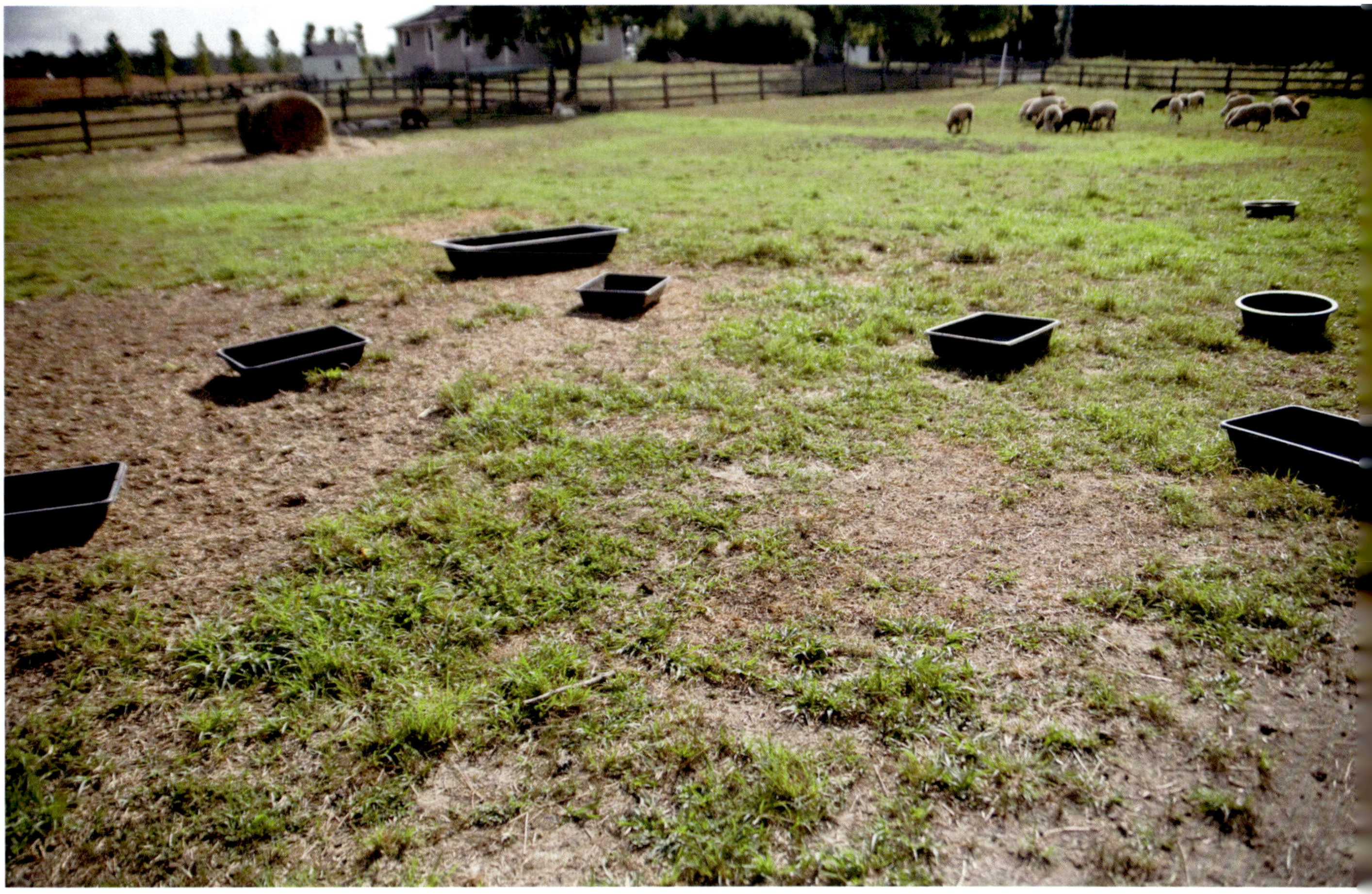

One of the many difficulties of raising sheep is finding someone to trim them every summer. Faith Farm took in these sheep from another farm after they had no more use for them and now the sheep roam the land freely.

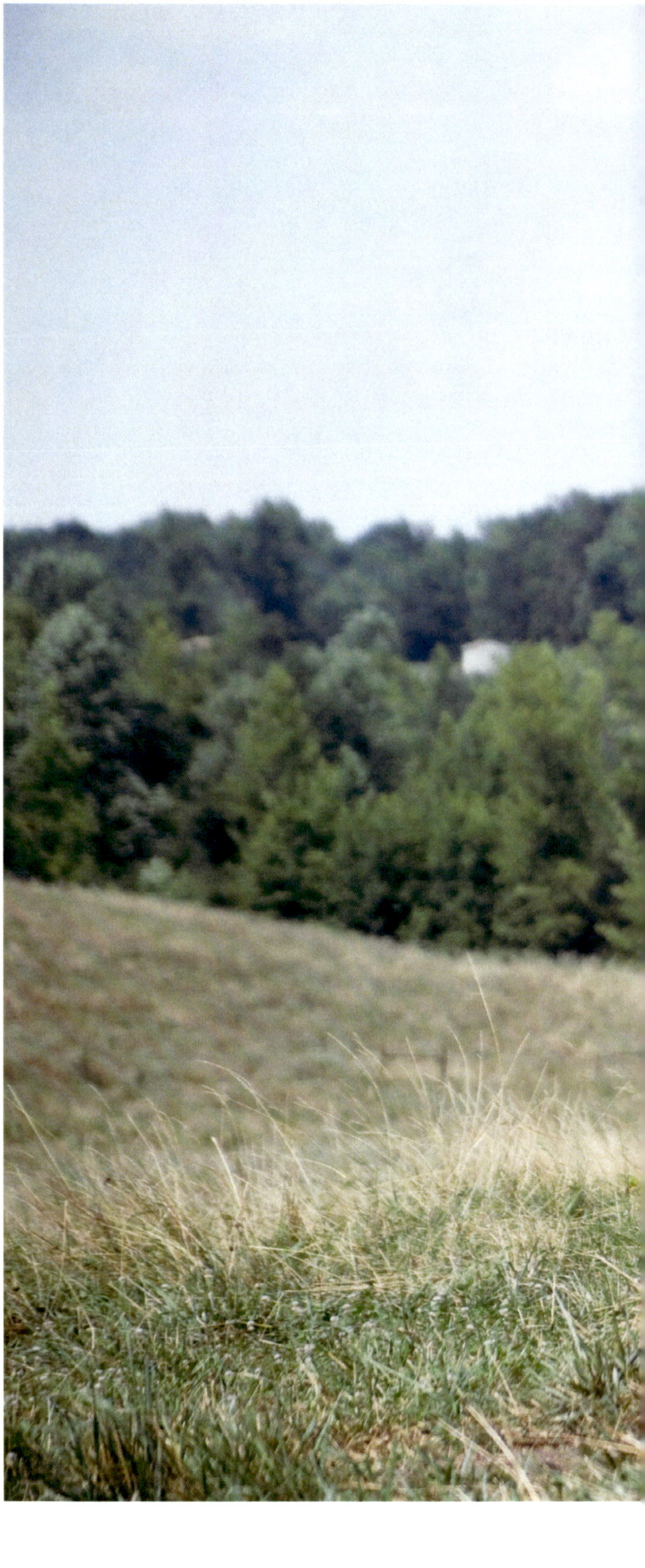

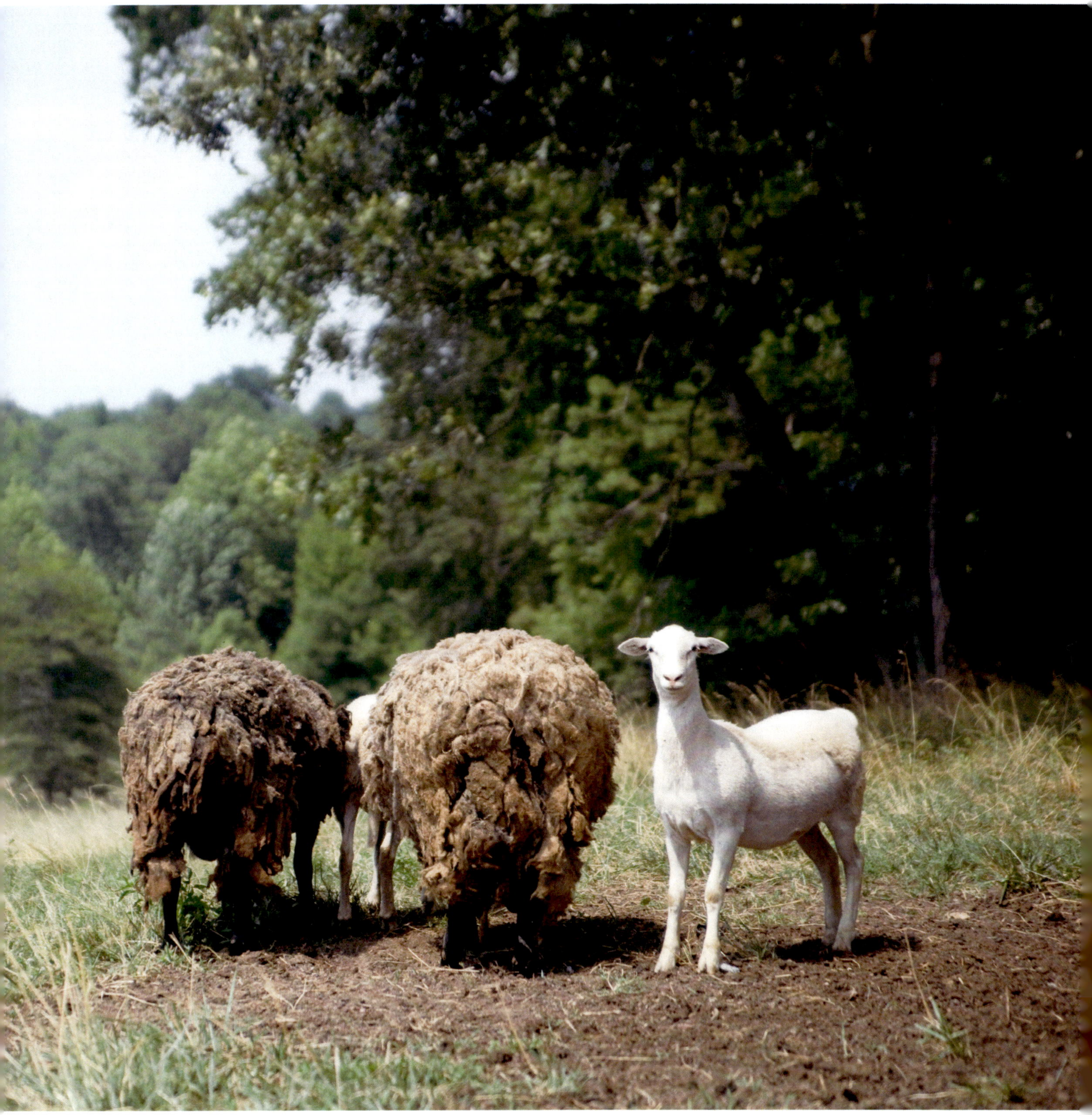

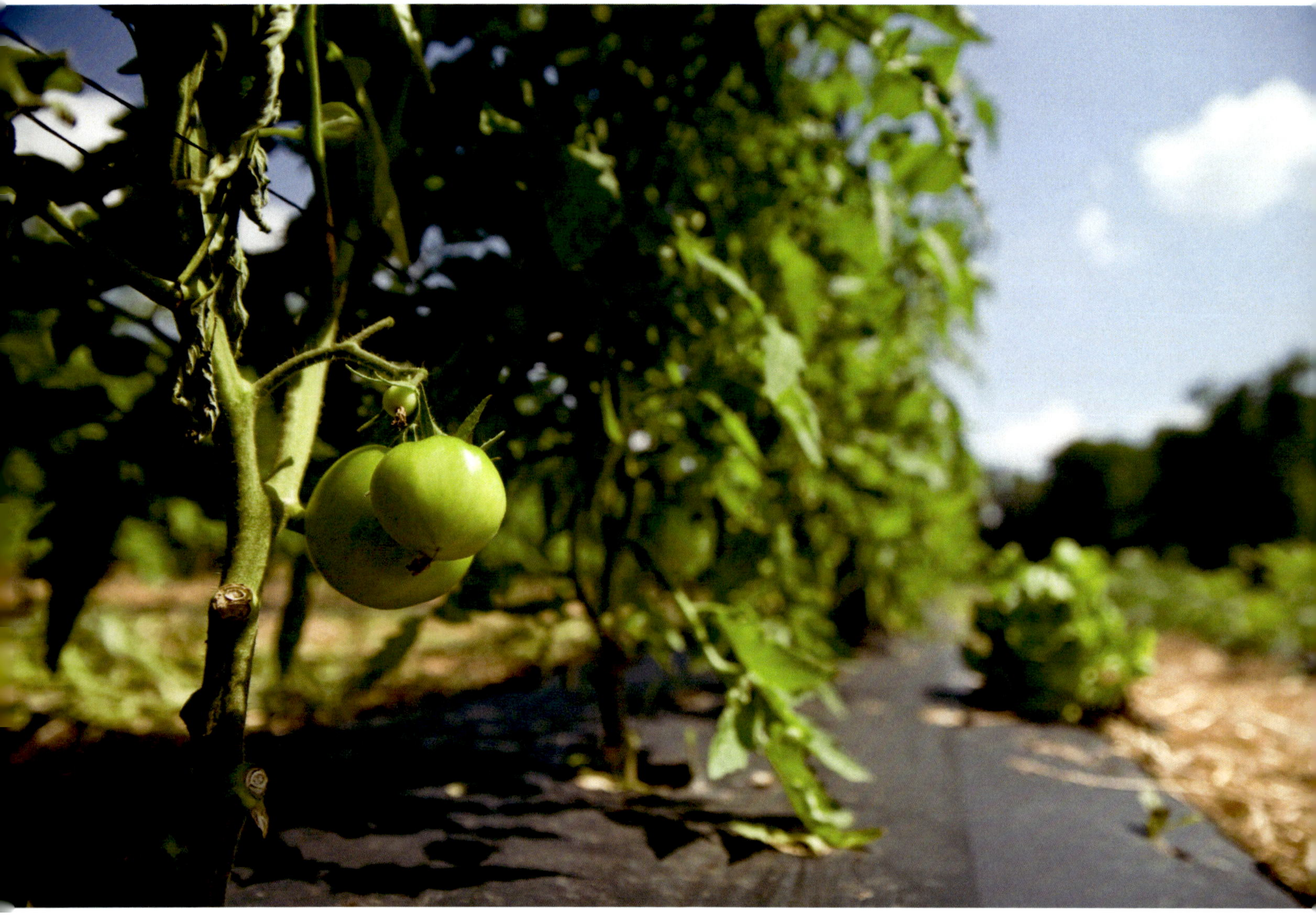

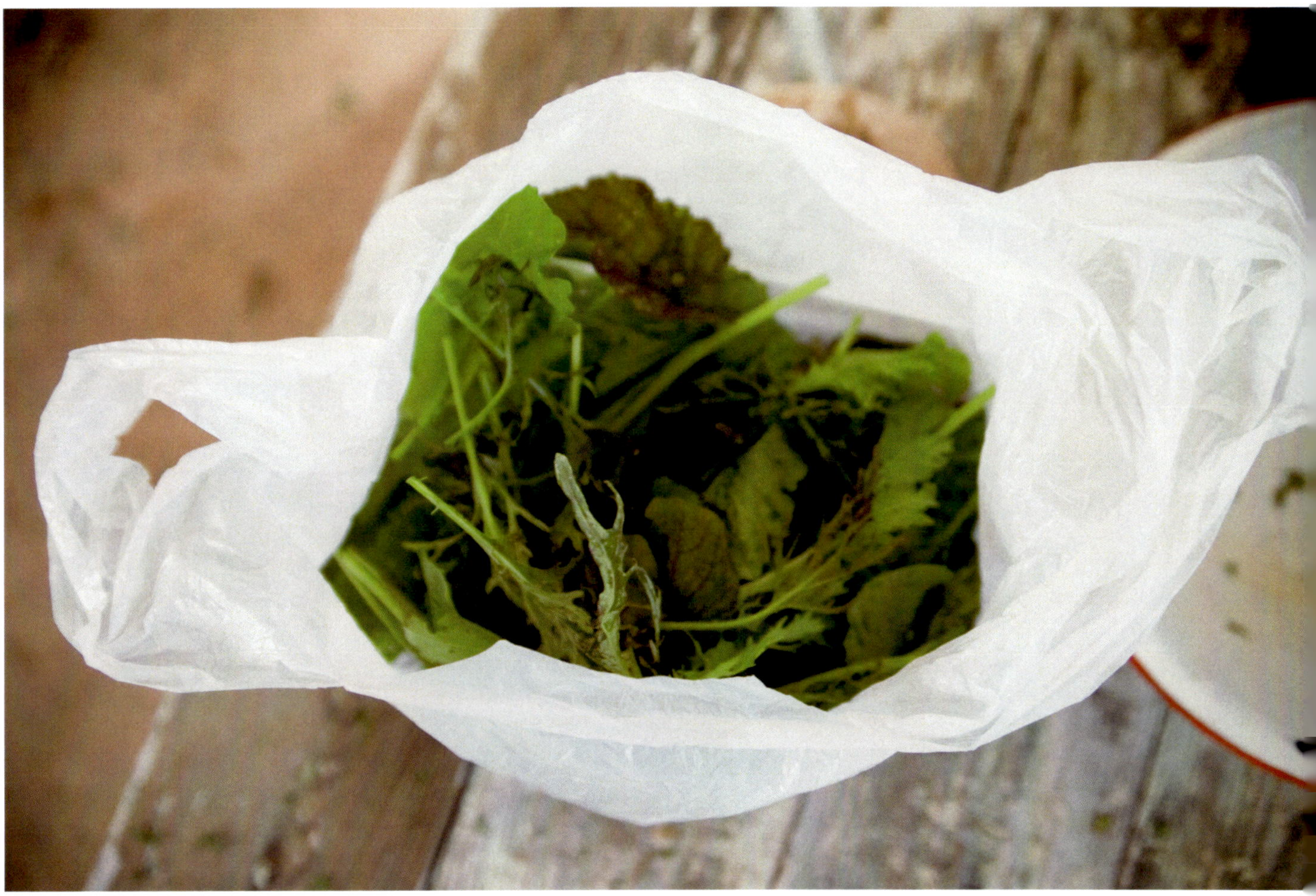

The namesake of Red Dog Farm relaxes at the Ashland Farmers Market—one of the most local markets in the sense that farmers travel the shortest distance to market. Many of the other Richmond markets attract vendors that travel one or two hours to sell produce.

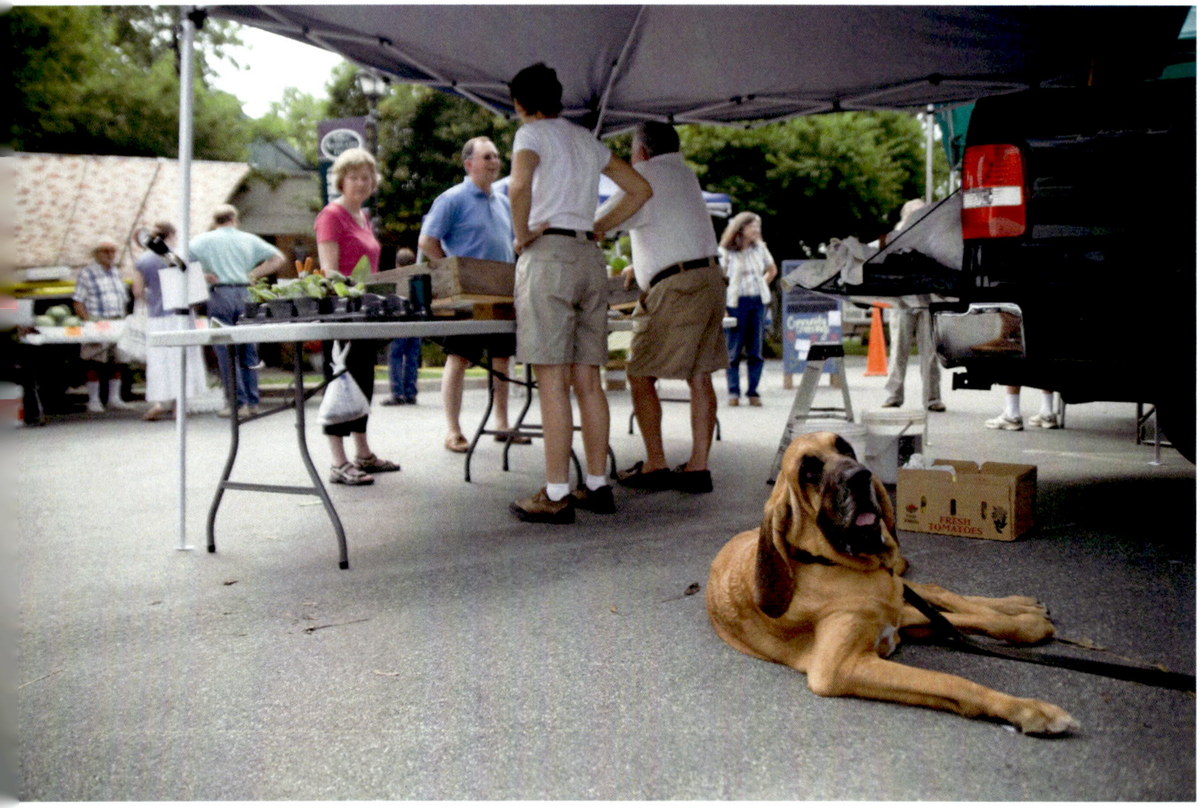

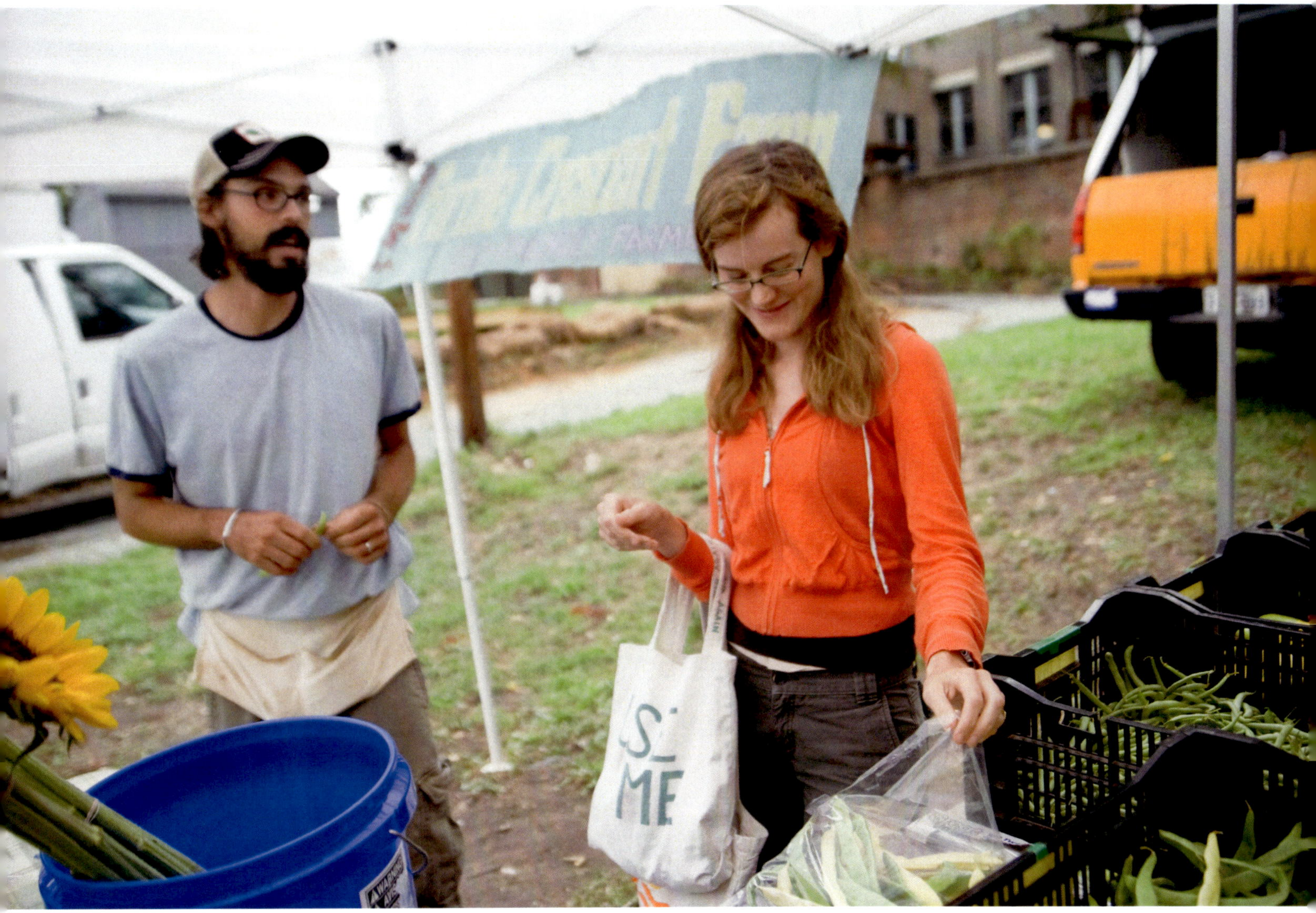

Picking up a CSA share at Fertile Crescent's Byrd House Market booth.

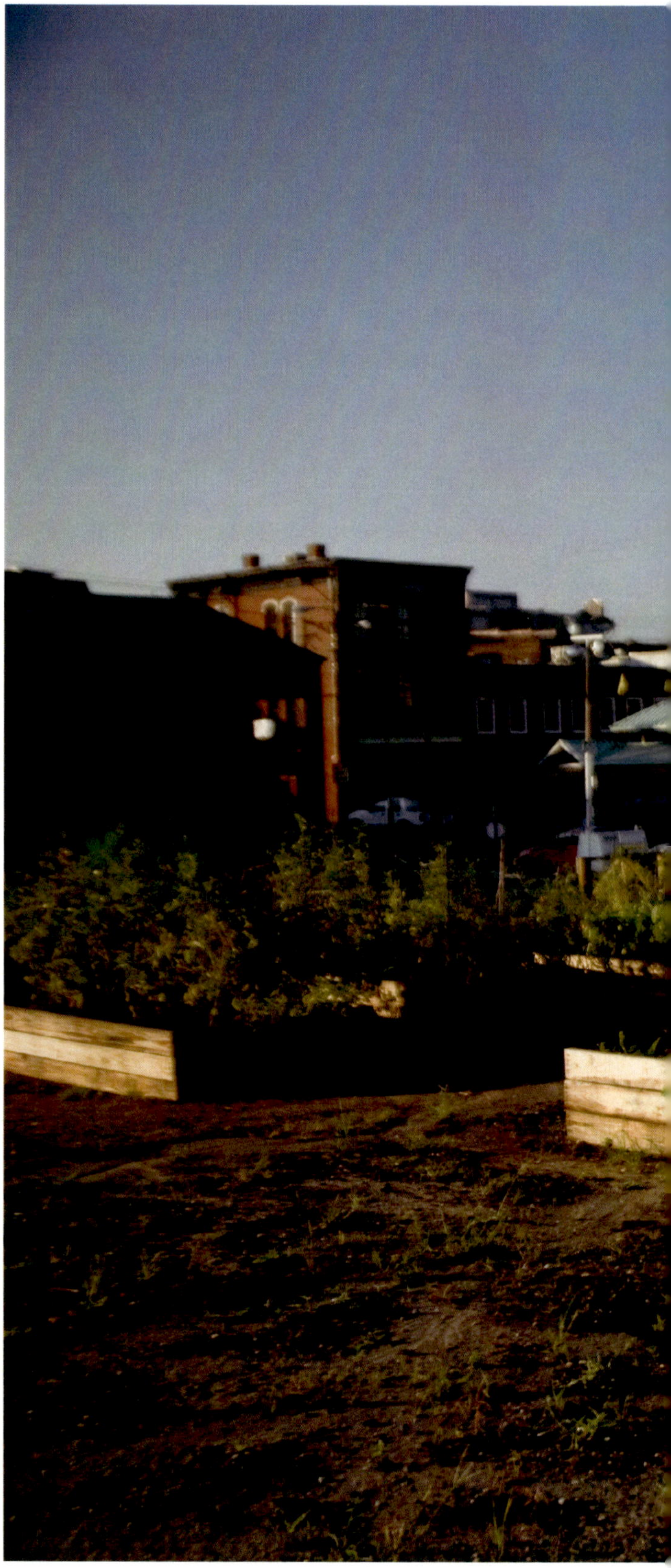

Richmond's 17th Street Farmers Market garden provides its produce to local food kitchens and restaurants in the neighborhood. The actual farmers market, on Thursdays, has had trouble maintaining farmers and buyers, losing them to more successful afternoon and weekend markets further from downtown.

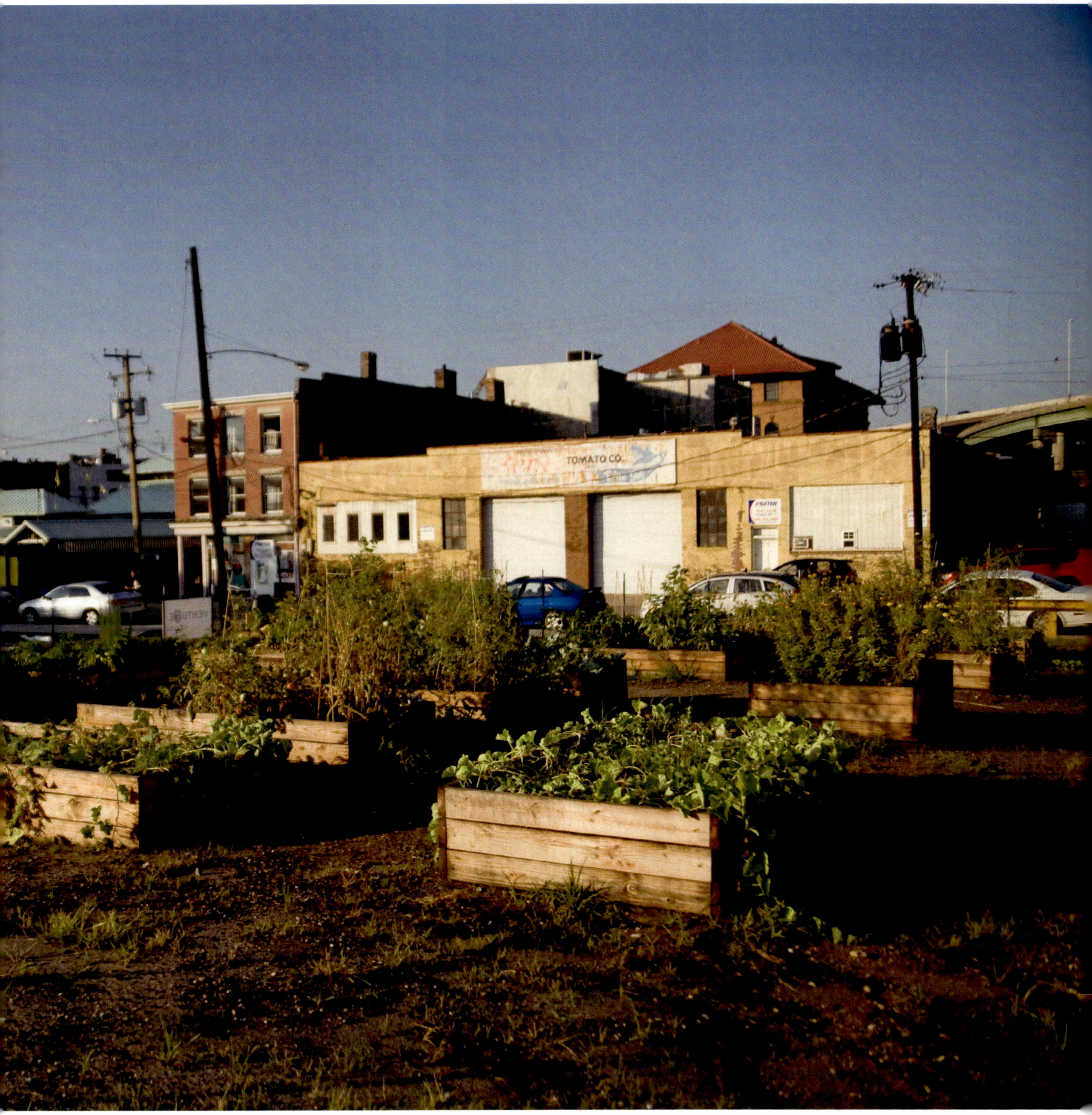
TOMATO CO.

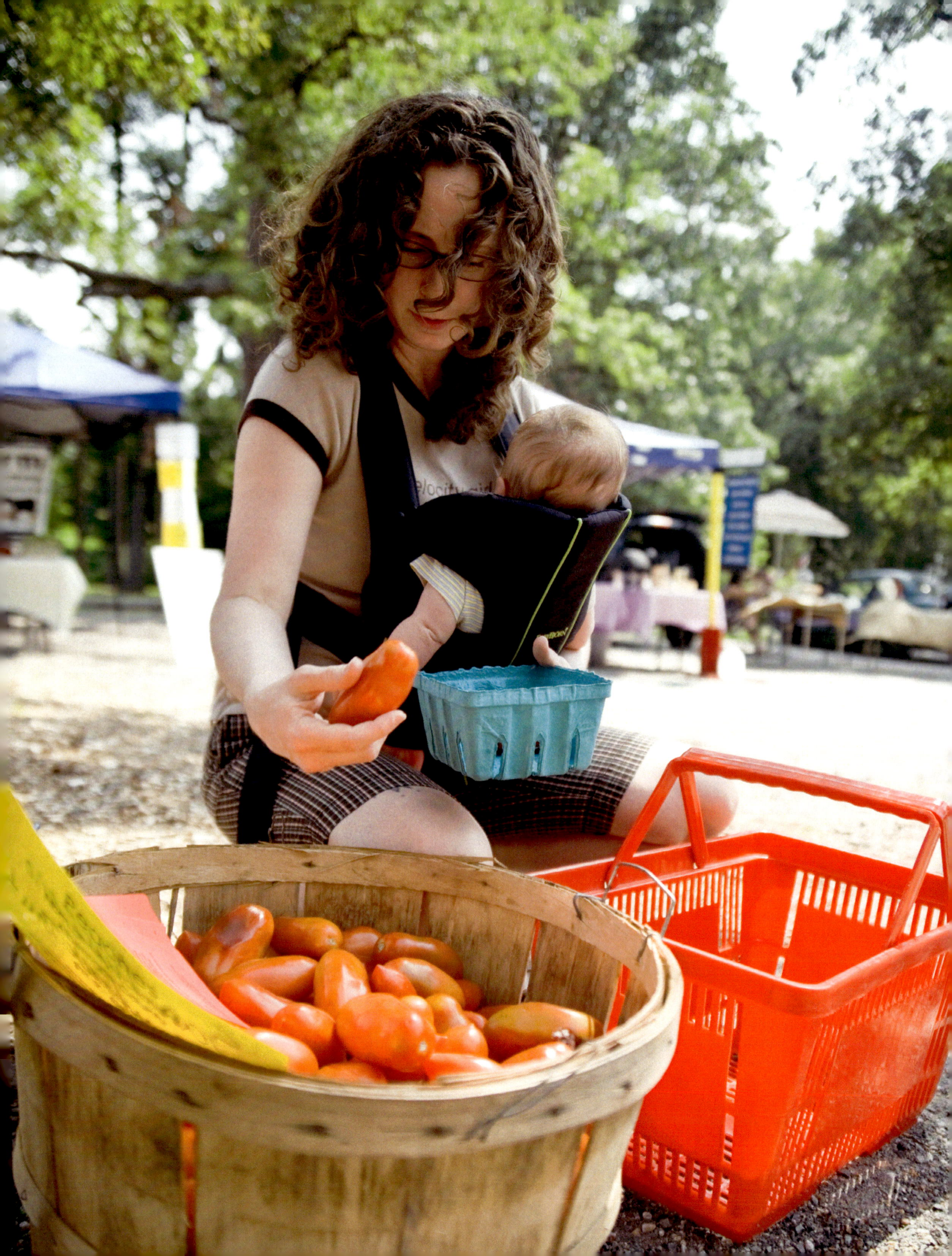

Kim and her son Ewan pick out tomatoes as part of their CSA share at Victory Farms. Victory visits several Richmond area farmers markets so that their 400 members can choose where to pick up their shares. Victory also invites their members to a CSA potluck every summer to see the farm and how all the produce is grown.

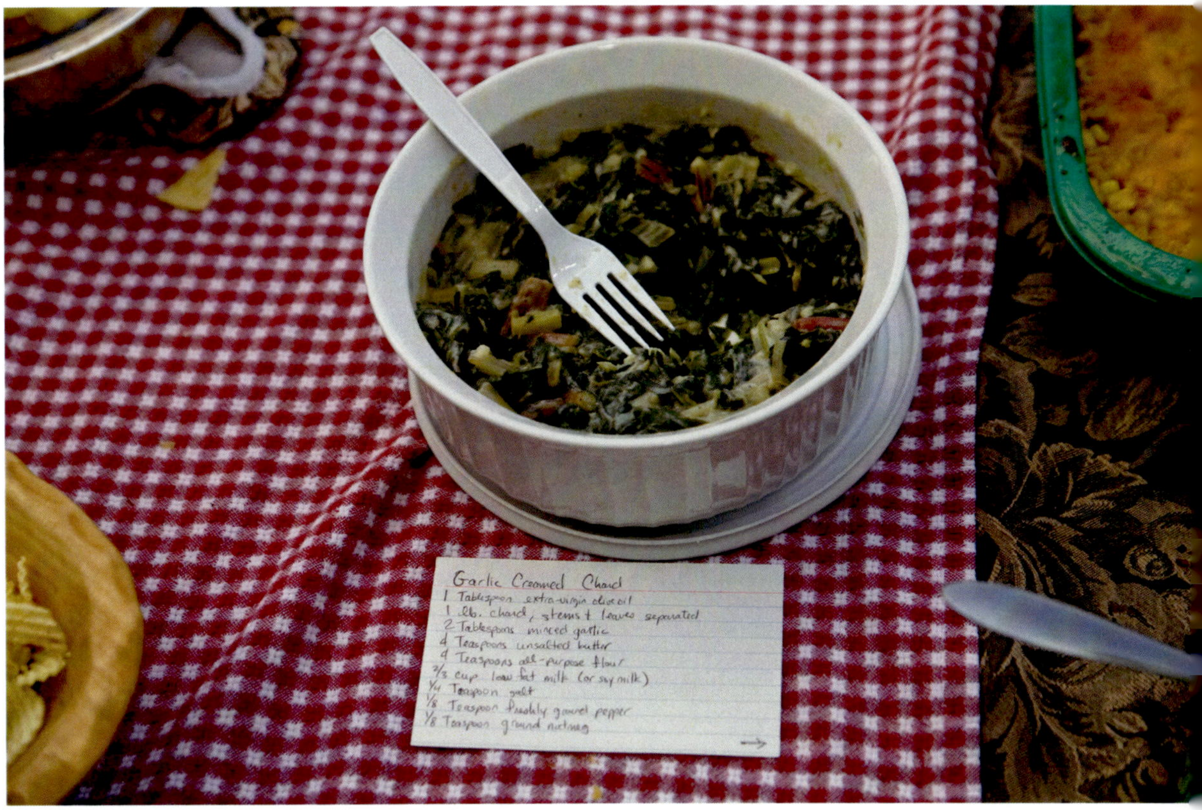

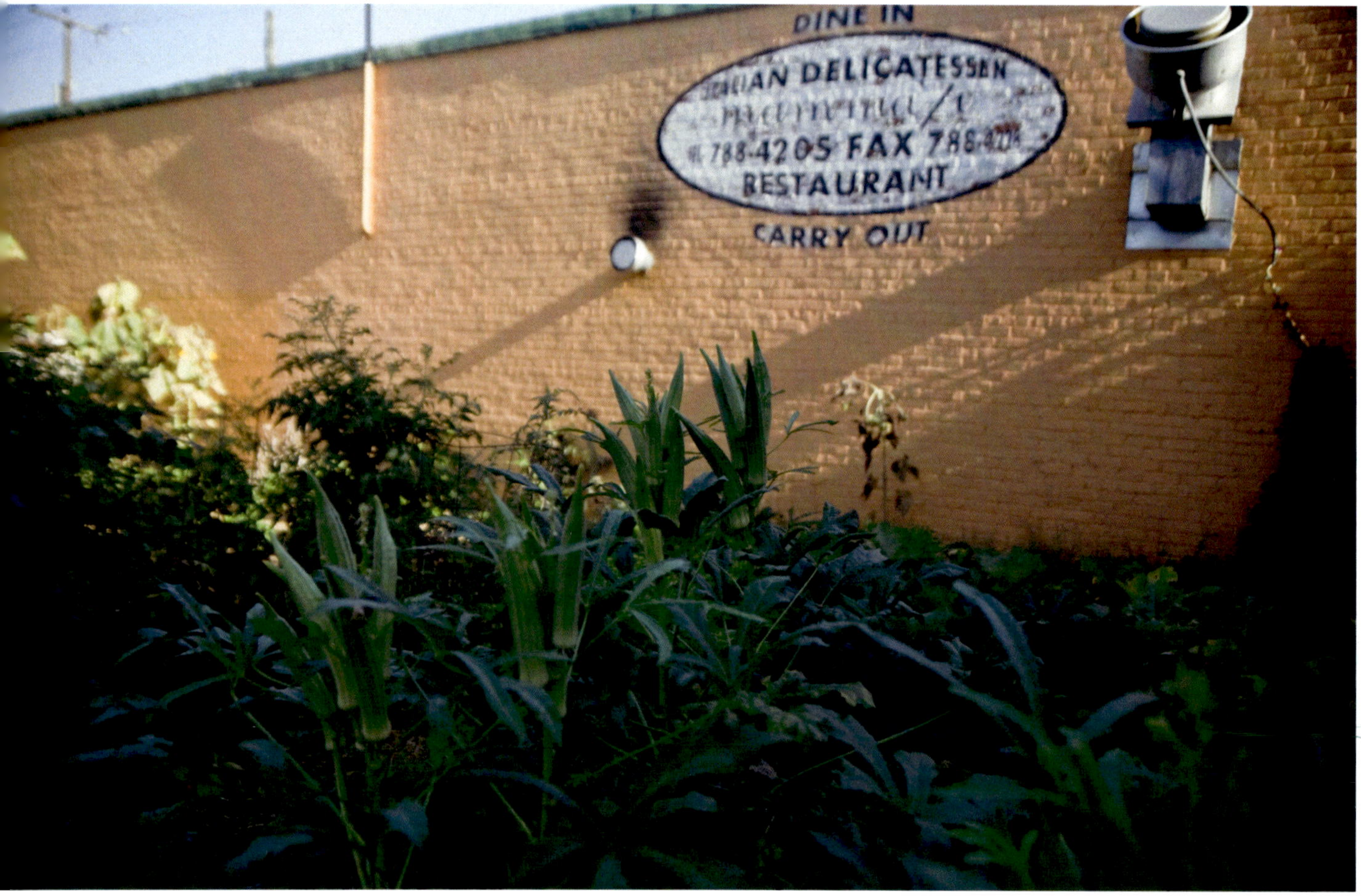

Mamma Zu, a popular Richmond restaurant, maintains its own garden right next door and uses as much produce from it as possible. Okra is pictured here, but basil and tomatoes are two other important crops for the Italian eatery.

Sub Rosa operates a bread subscription service, where you pay monthly for a loaf of bread a week and they provide pickup at area farmers markets. Frustrated by the lack of local, sustainable grains, the owners spearheaded an initiative to grow wheat locally and sustainably, but many of the first test plots of wheat drowned because of heavy spring rains—demonstrating both the difficulty of growing some crops and how weather-dependent crop production can be.

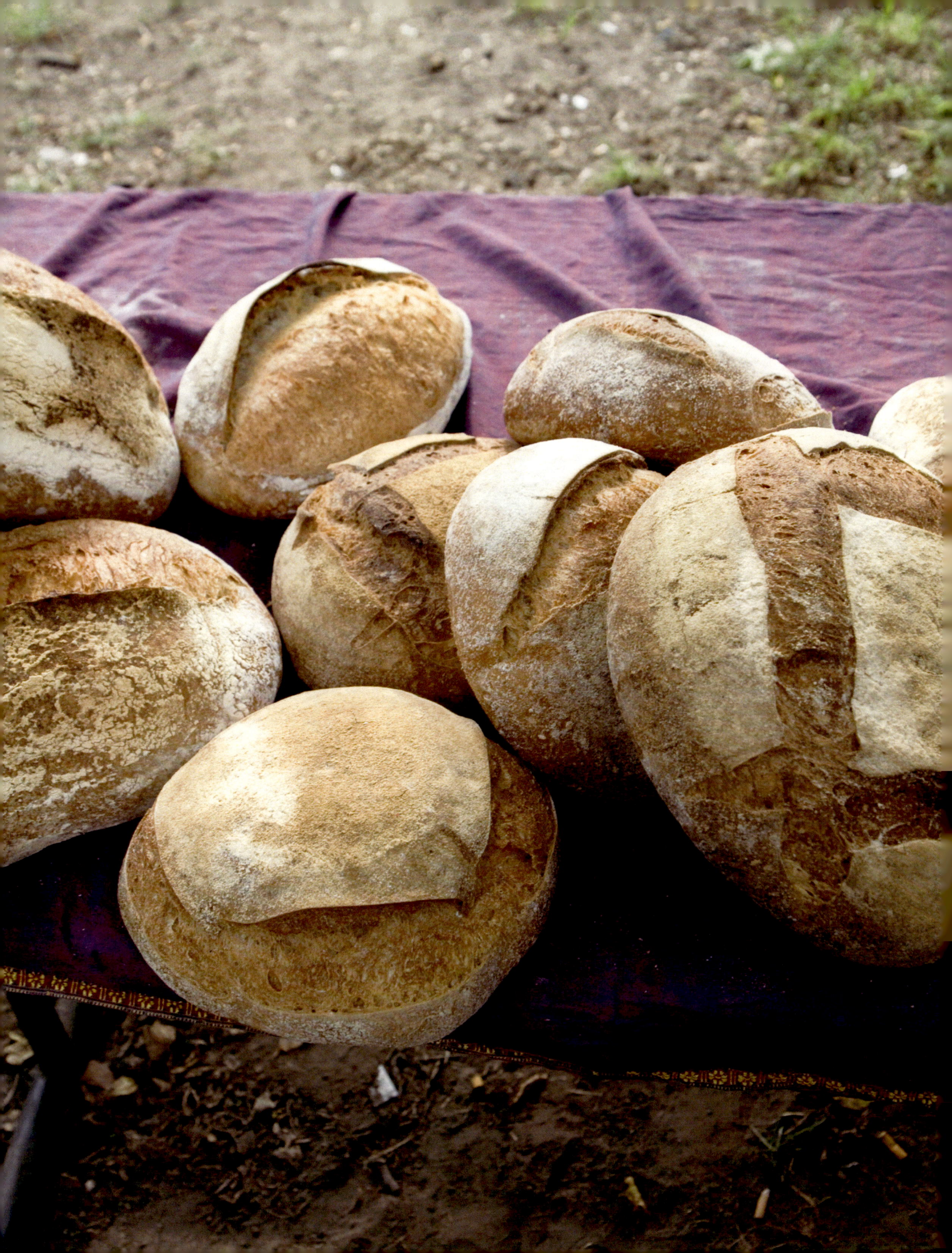

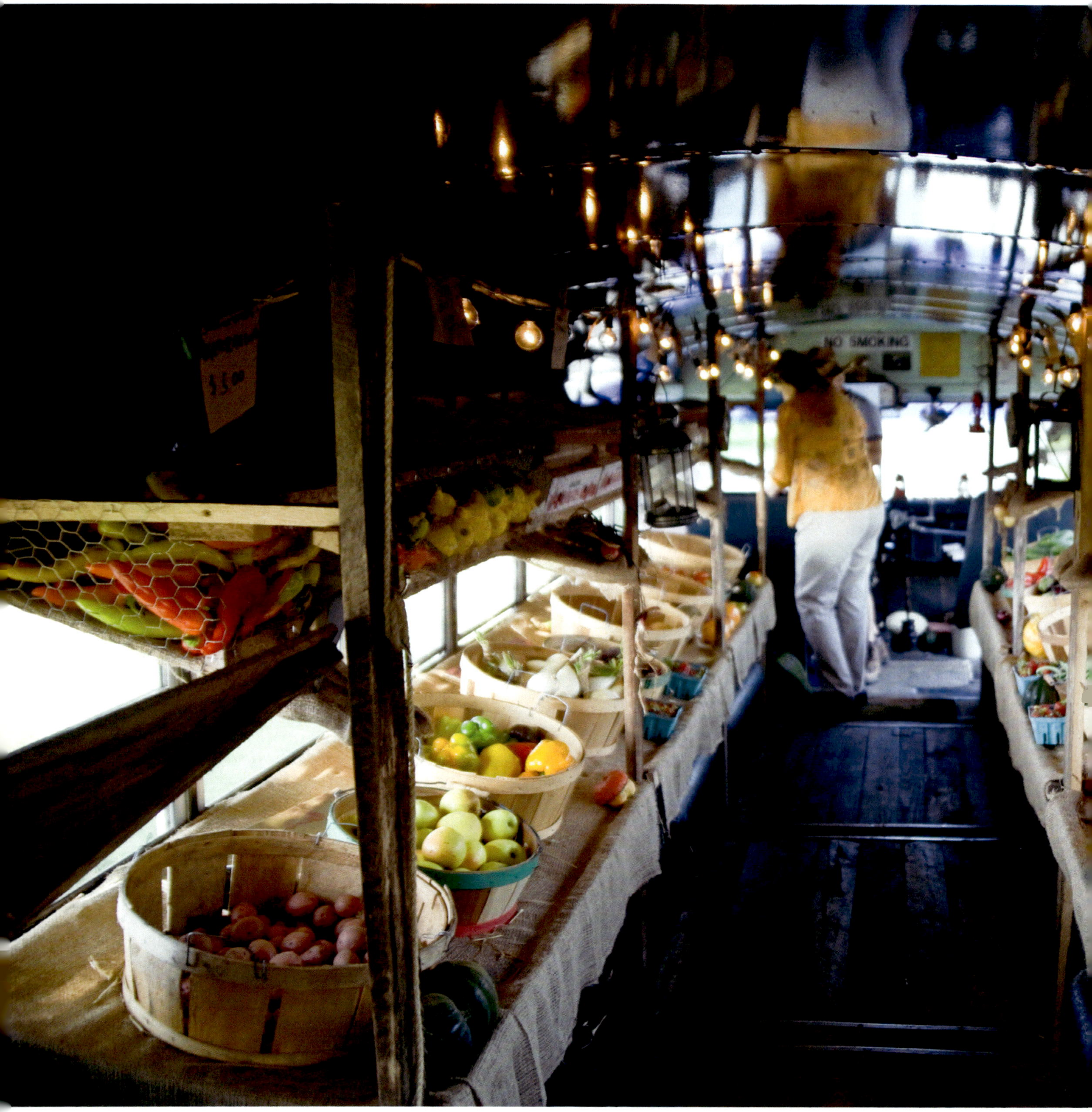
NO SMOKING

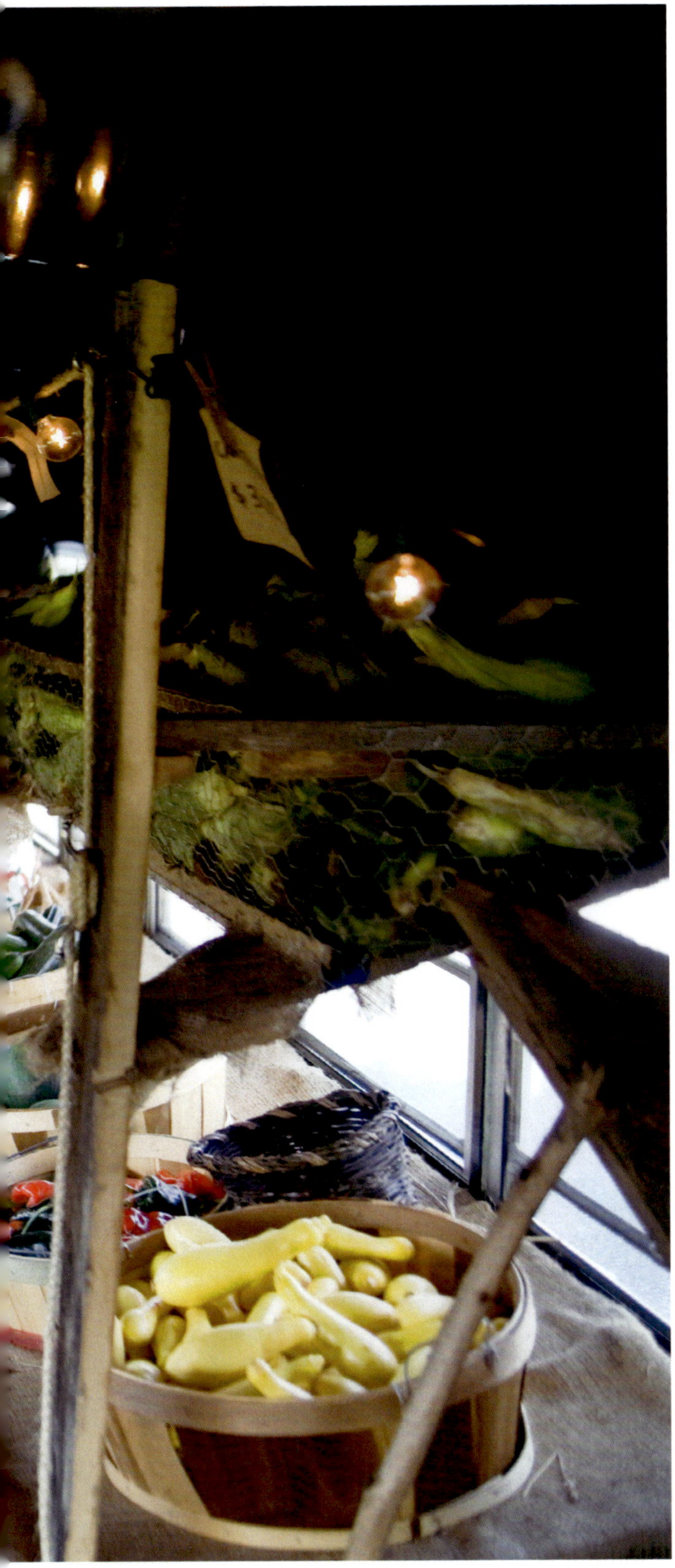

Farm To Family, started in 2009 in a used school bus, provides a mobile market of local produce all year. The best part about the bus is it can visit food-scarce neighborhoods—nowhere near grocery stores—and pick up food directly from the farmers. The idea behind the bus was to create more food security through better access to local food while educating Richmonders about their food.

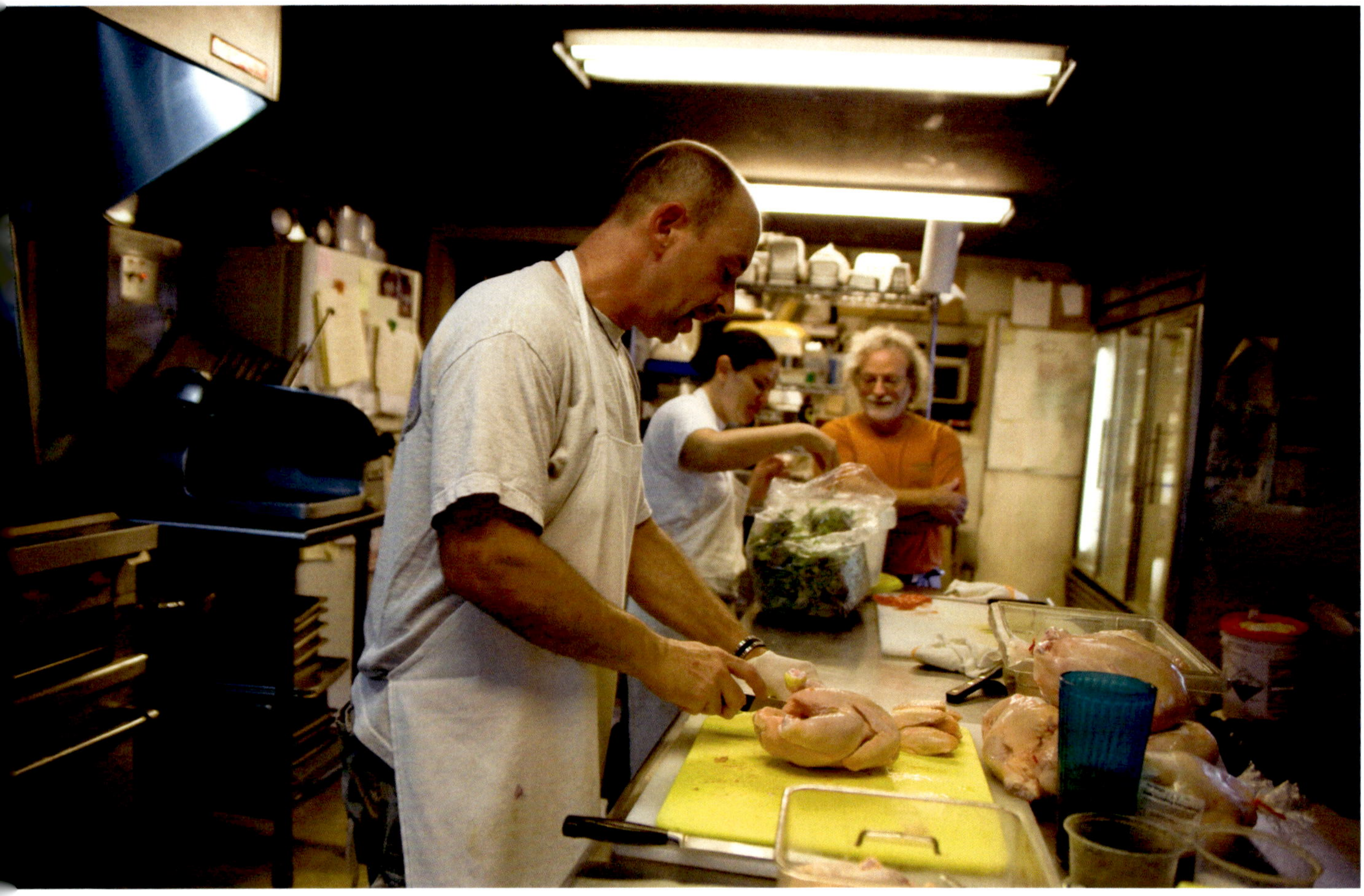

Dave and Dee deliver produce as part of the Fall Line Co-op to Edible Gardens restaurant, which only served local food. The restaurant has since gone out of business, but the co-op continues and has several drop off points throughout Richmond.

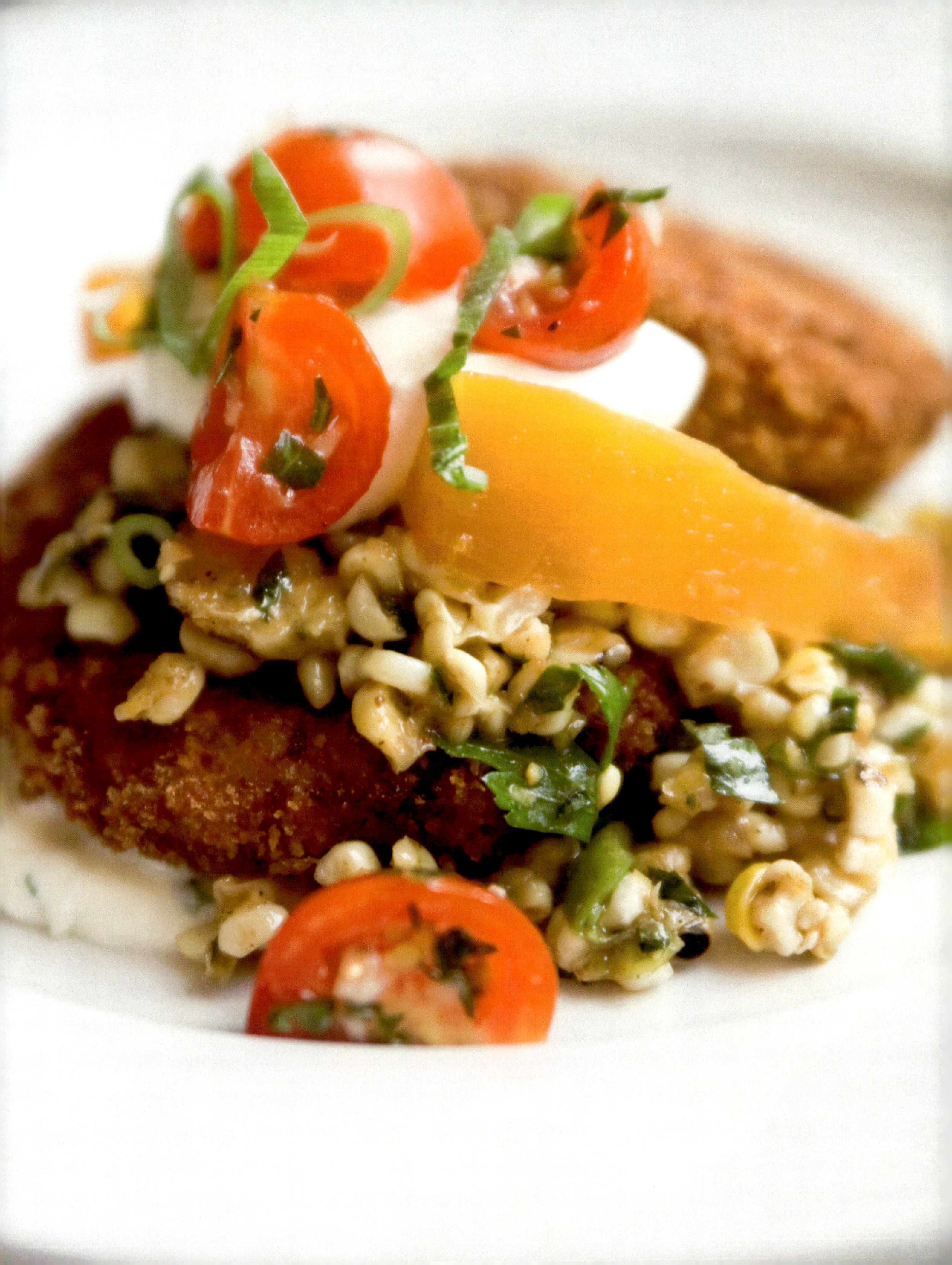

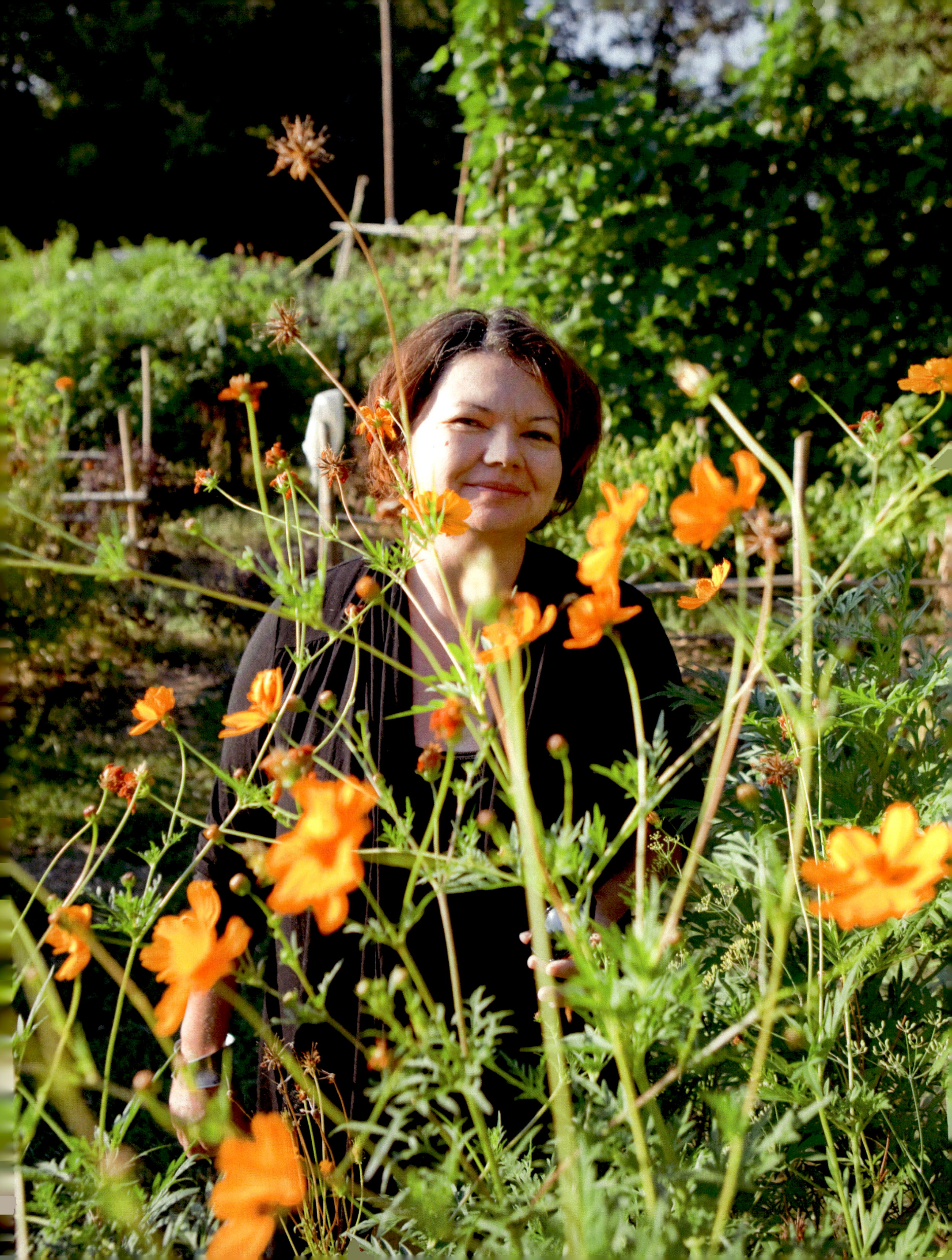

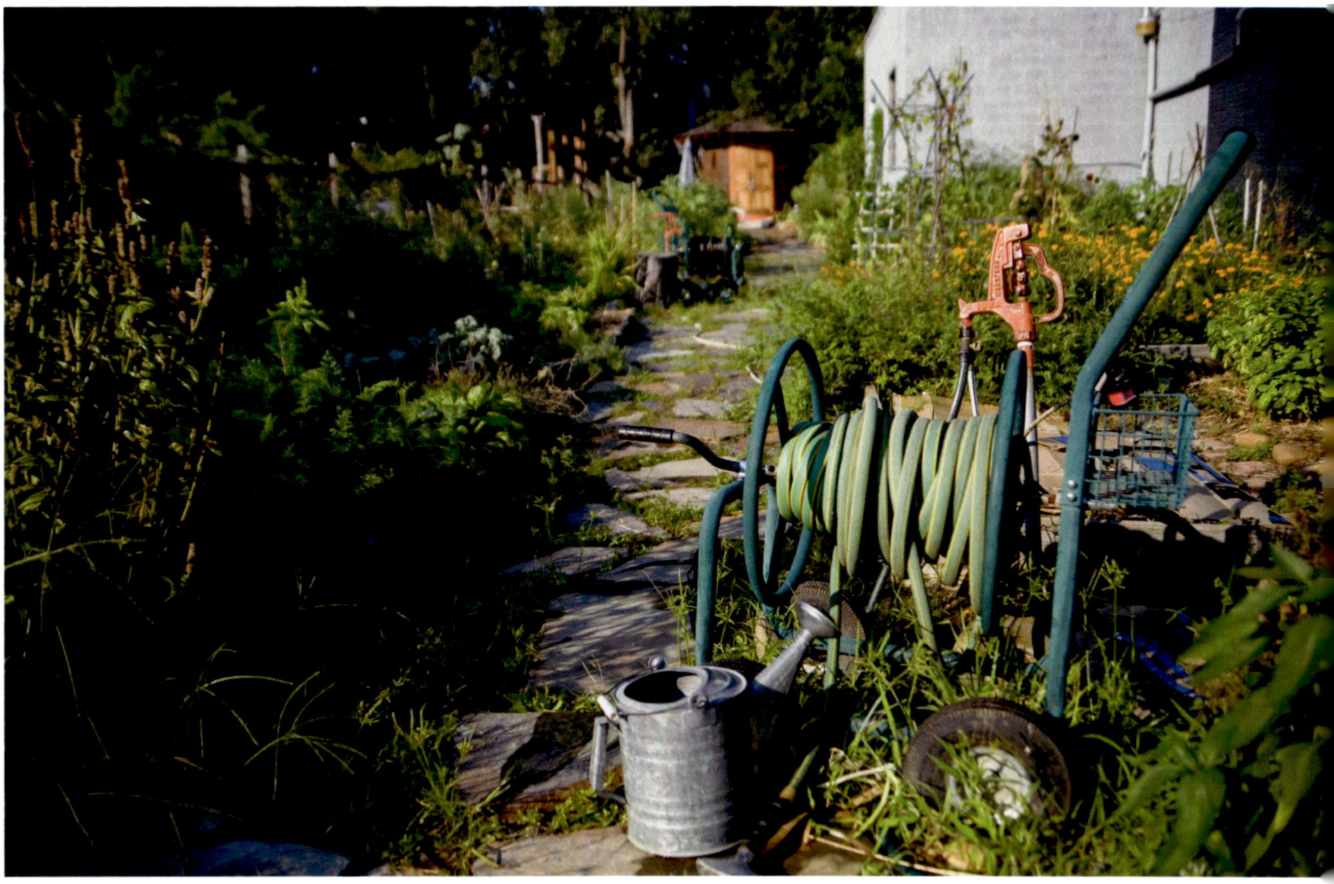

Lisa Taranto is the former director of Tricycle Gardens and new president of the Allegheny Mountain School.

The first community garden started by Tricycle Gardens is on Jefferson Avenue in Richmond. Tricycle Gardens is expanding its community gardens to include an urban farm, market, and a permaculture school in Western Virginia near Monterey. The goal of the nonprofit is to encourage sustainable living through healthy, available, and affordable food.

Over 200 volunteers, mostly from Home Depot, help construct a new community garden run by Tricycle Gardens in Chimborazo Park. The garden was built by reusing as much wood as possible from an old playground, including the railroad ties pictured to build the raised beds.

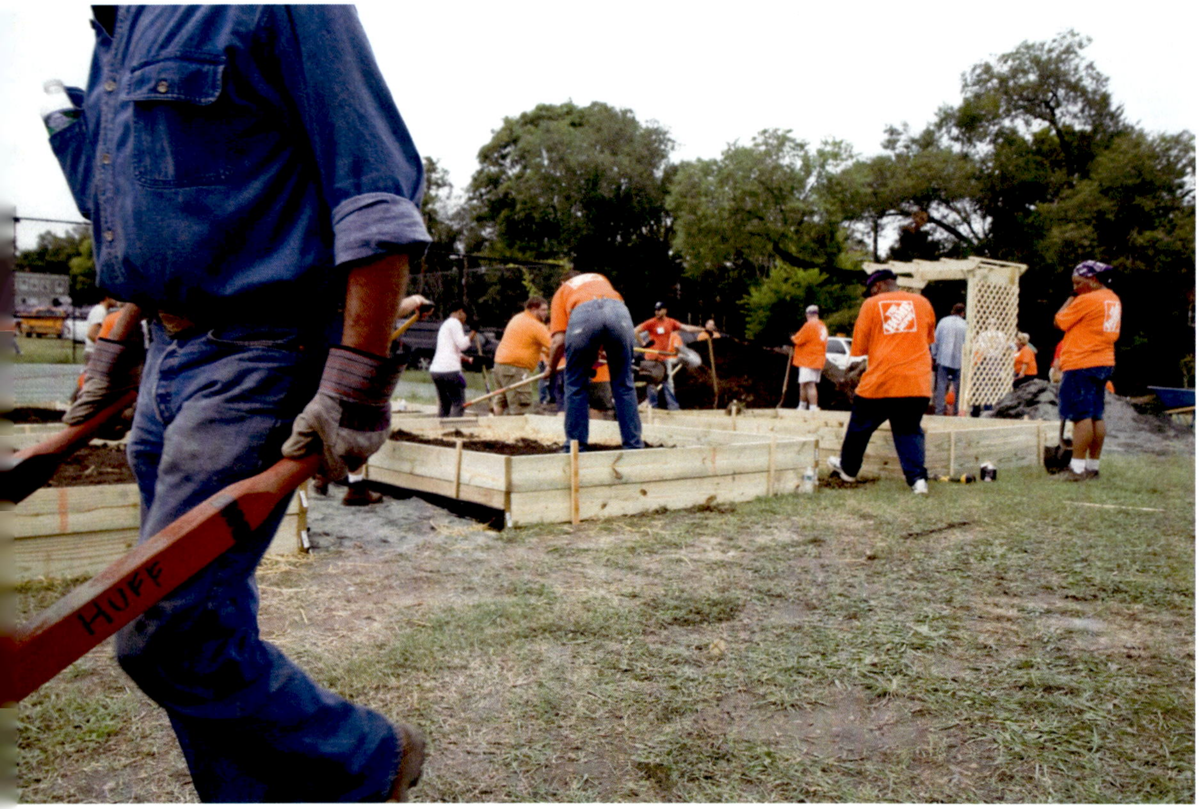

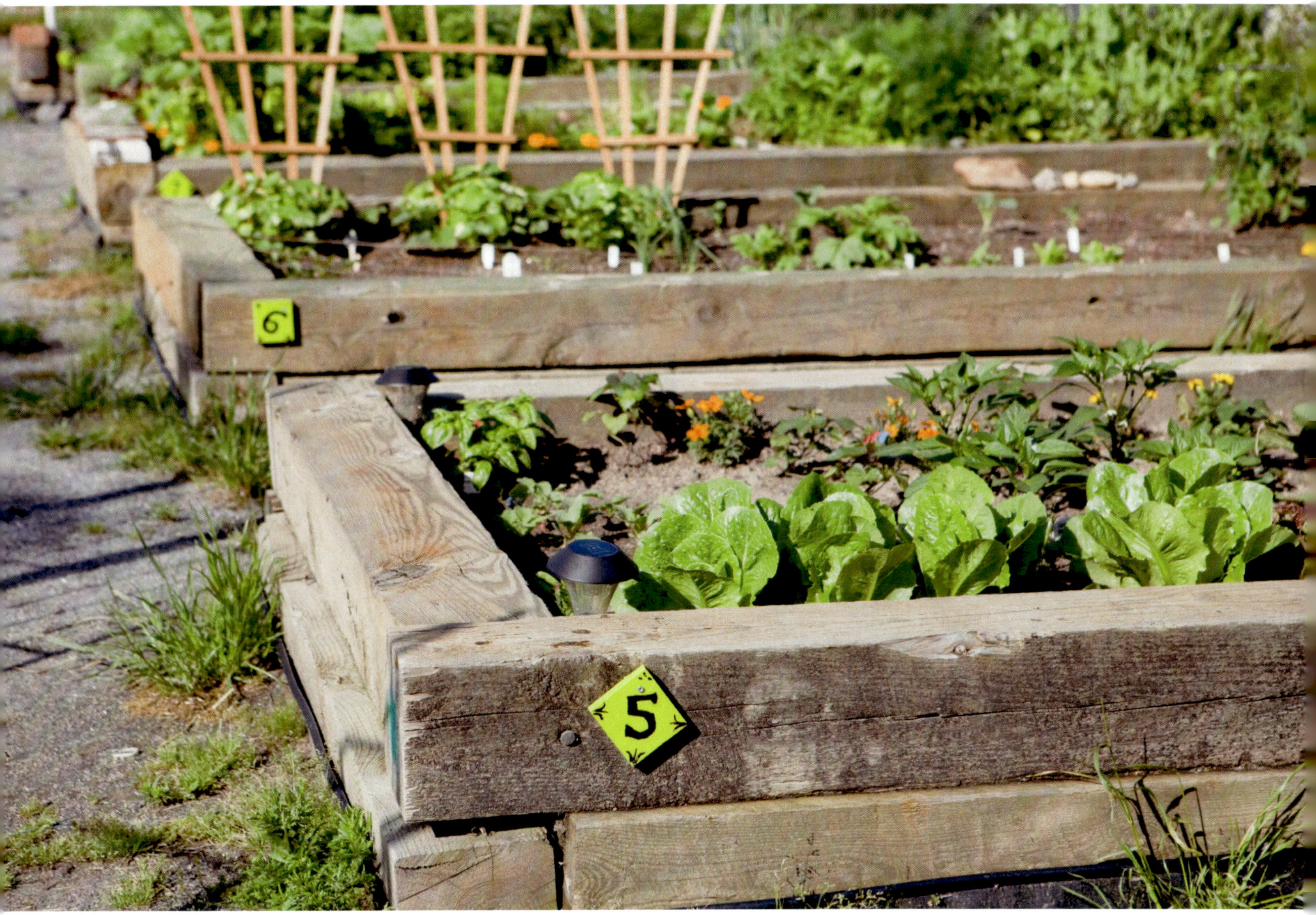
6
5

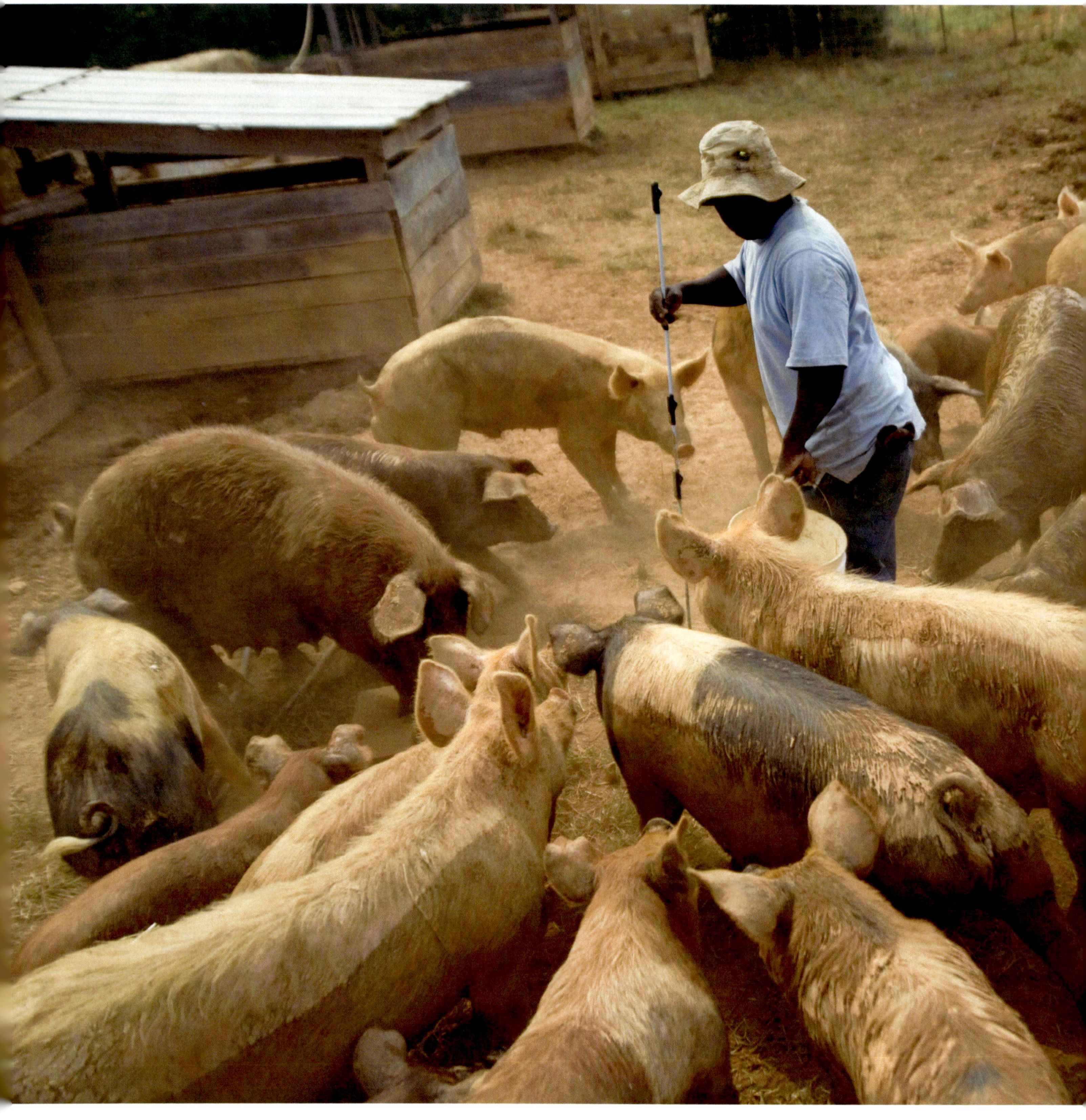

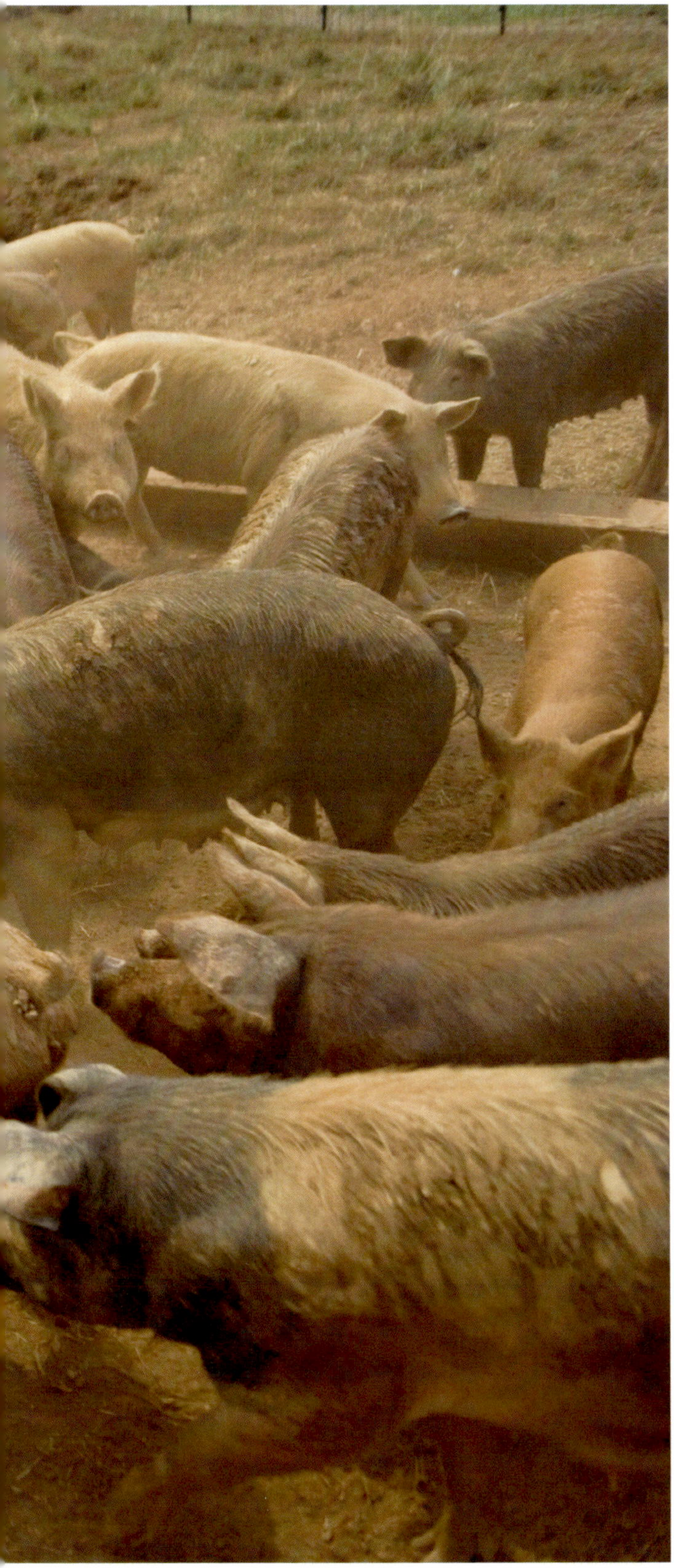

Feeding pigs at Faith Farm. Pigs can get aggressive when it is feeding time, so the worker has to be careful to stay on his feet and control the pigs so that they get their food and he remains safe.

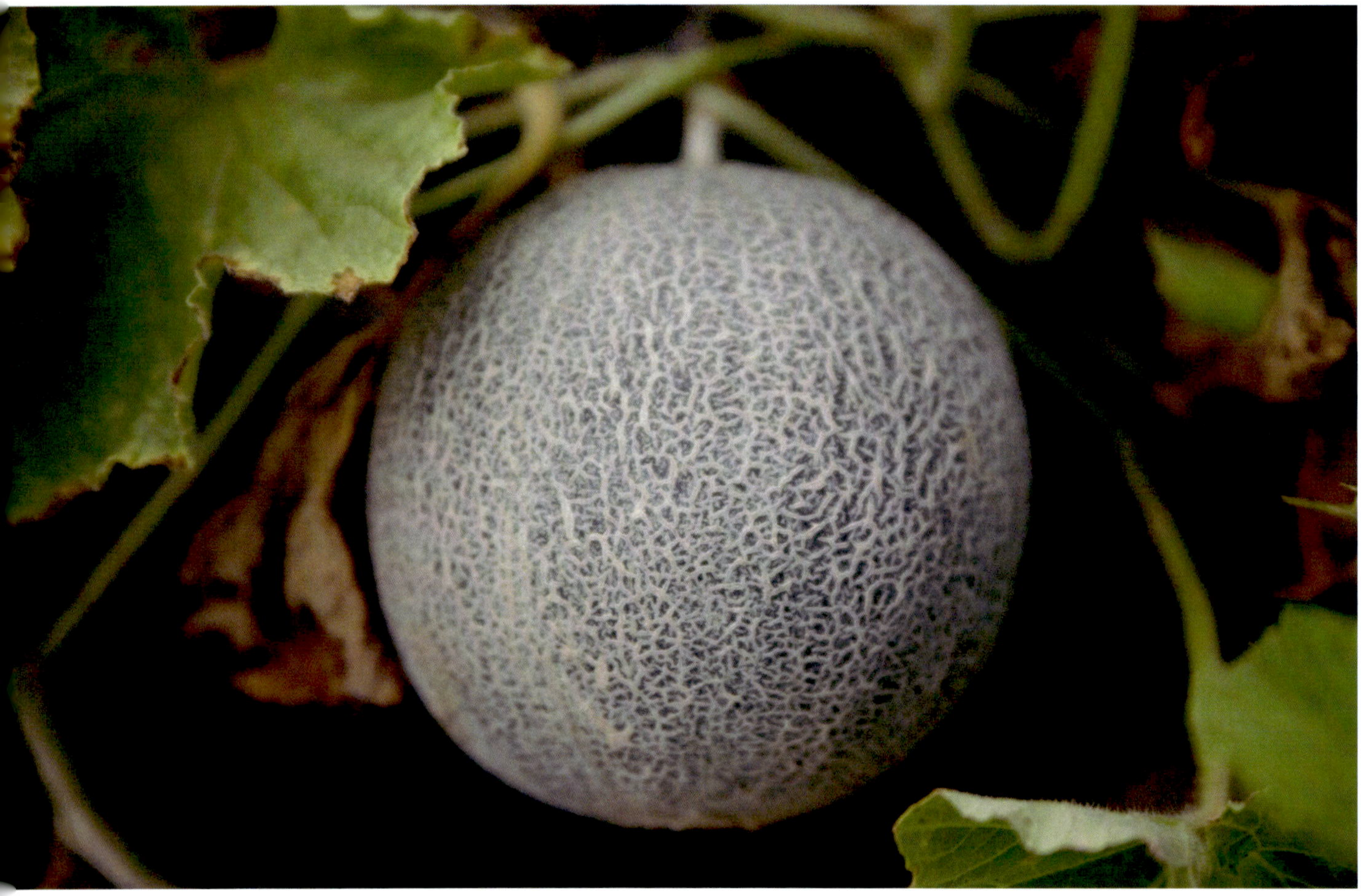

Above, a cantaloupe waiting to be picked at Trail's End Farm.

To the right, Dayspring Farm's squash harvest is separated for CSA shares, markets, and restaurant deliveries.

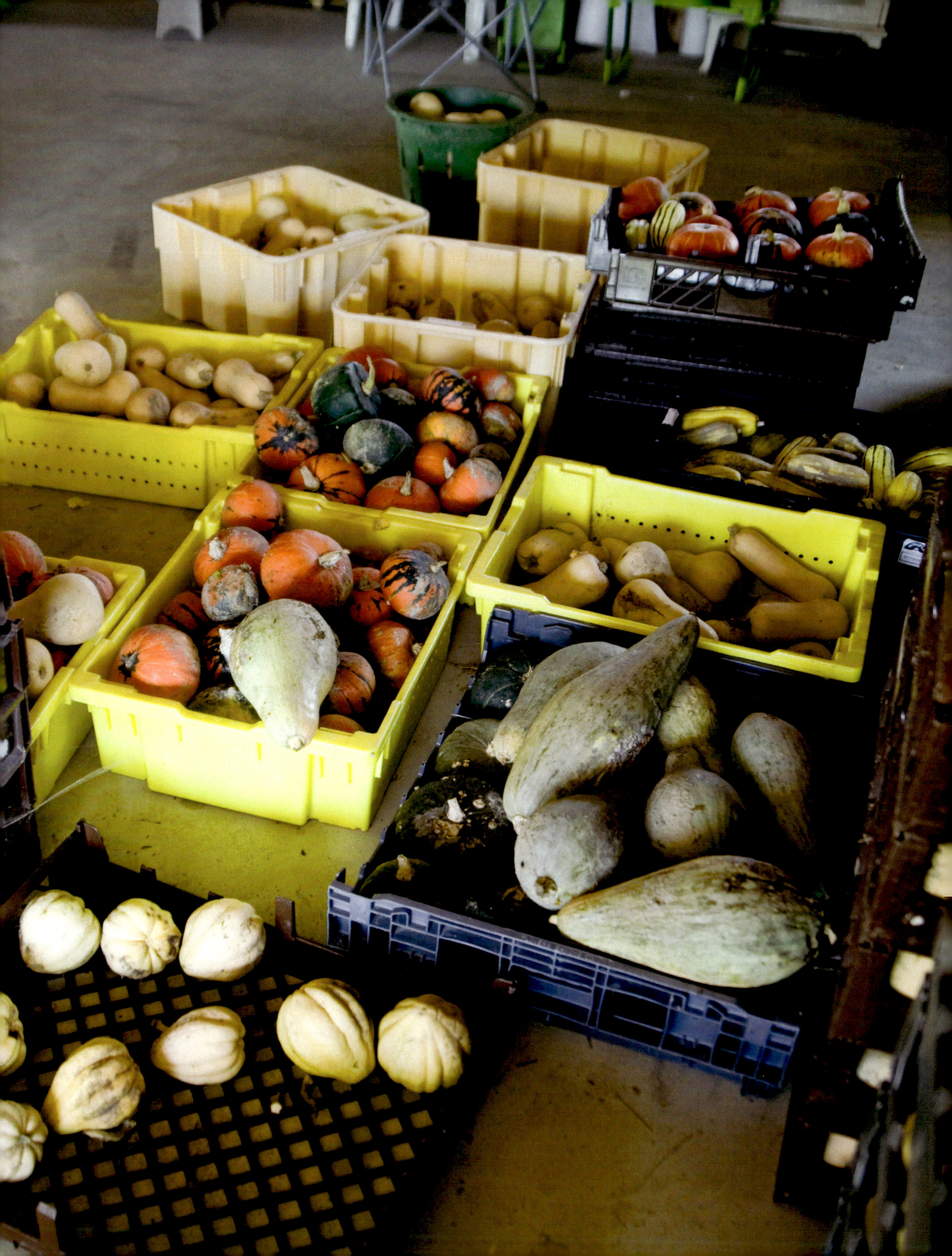

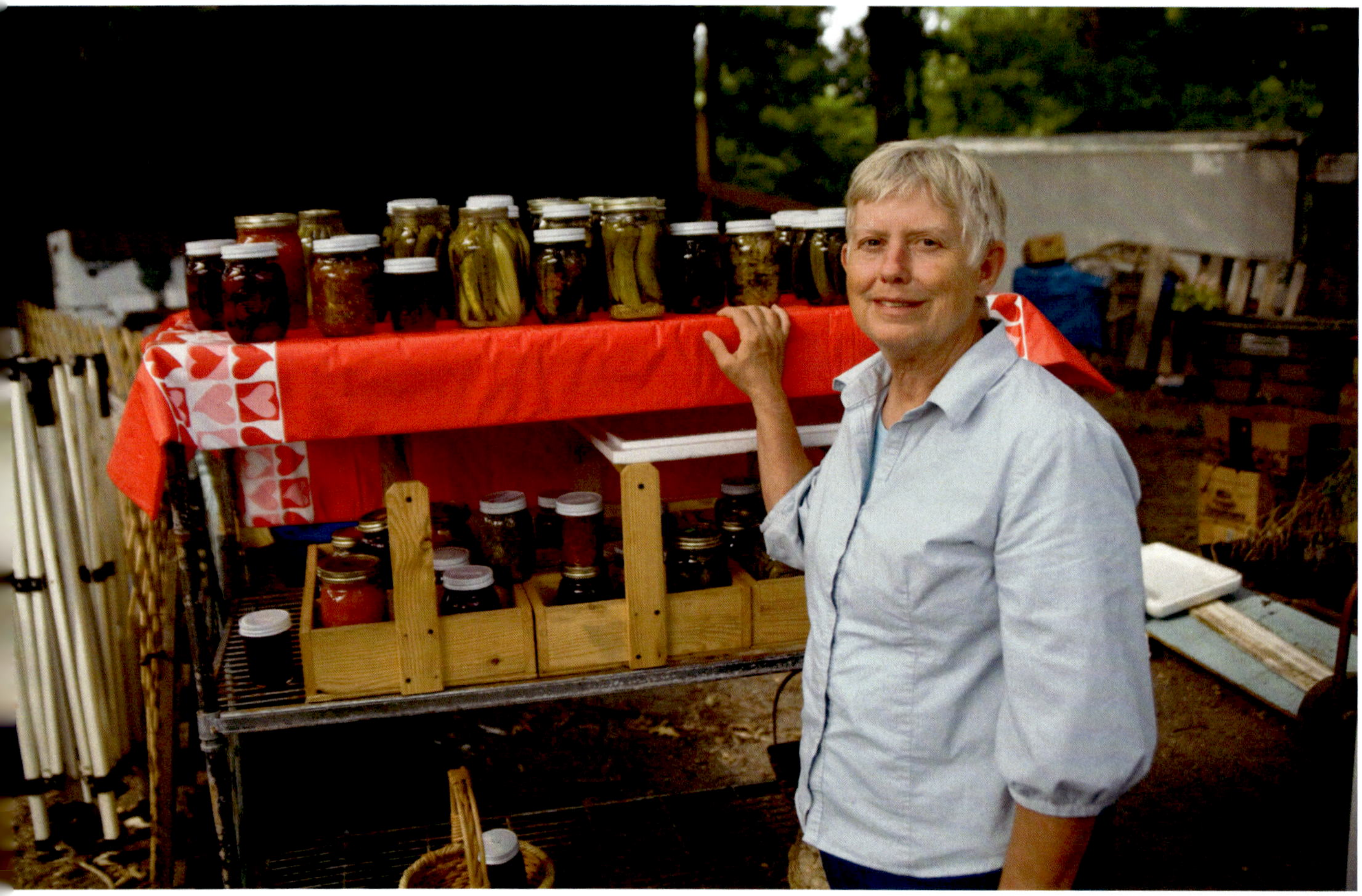

The Green Bay Open Air Market, owned by Carolyn, is one of the only shops in Green Bay, an hour and a half southwest of Richmond. The market offers local produce from farms and prepared goods from the local Amish community.

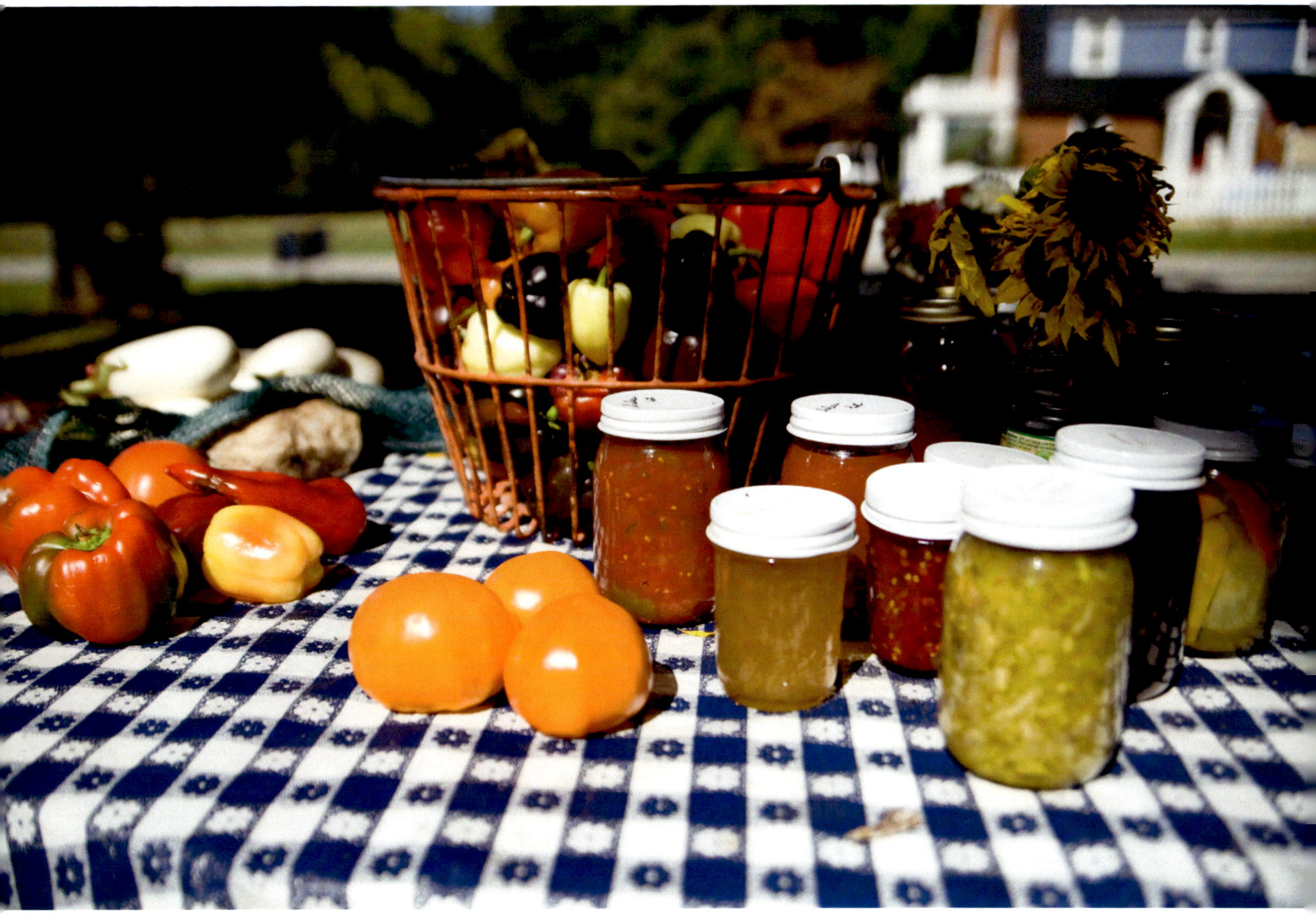

Anne makes jam at the Hanover County Cannery near Richmond. Having the local cannery allows both farms and local residents to preserve their produce to enjoy all winter. Increasing local preserving and canning is one important way to keep food local and sustainable.

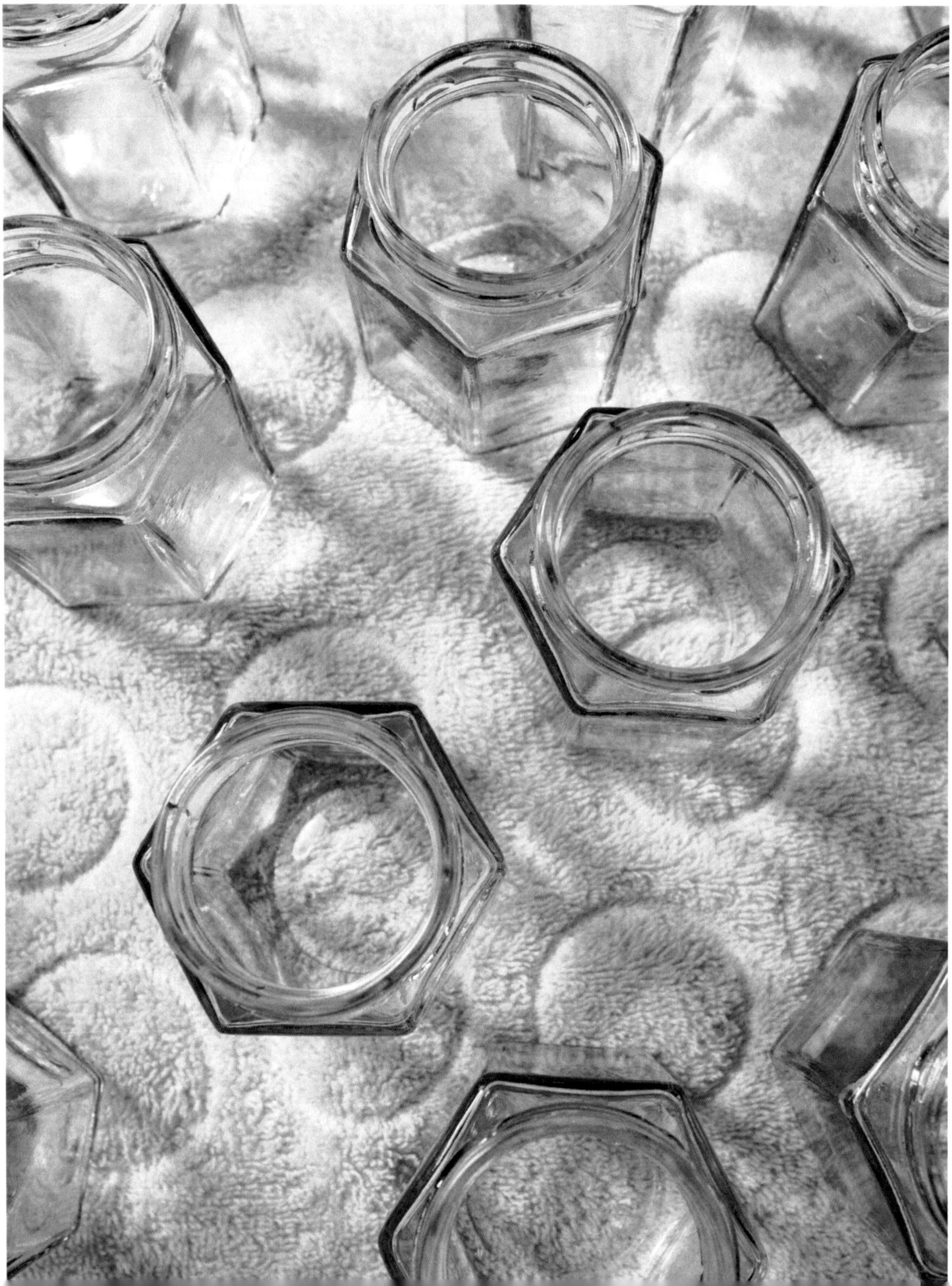

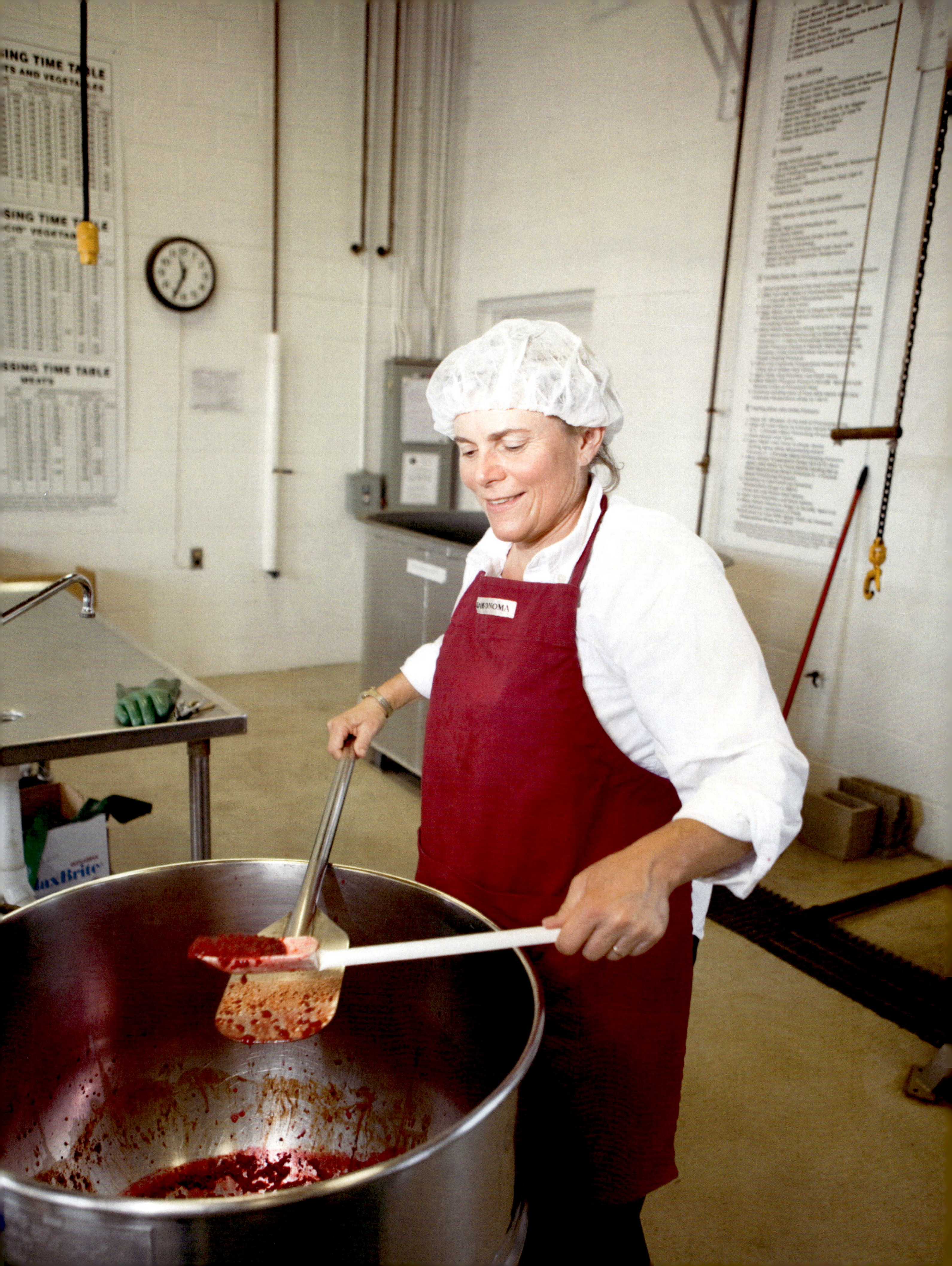

Eggplant from Susannah and Chandler's backyard garden thawing for a winter meal.

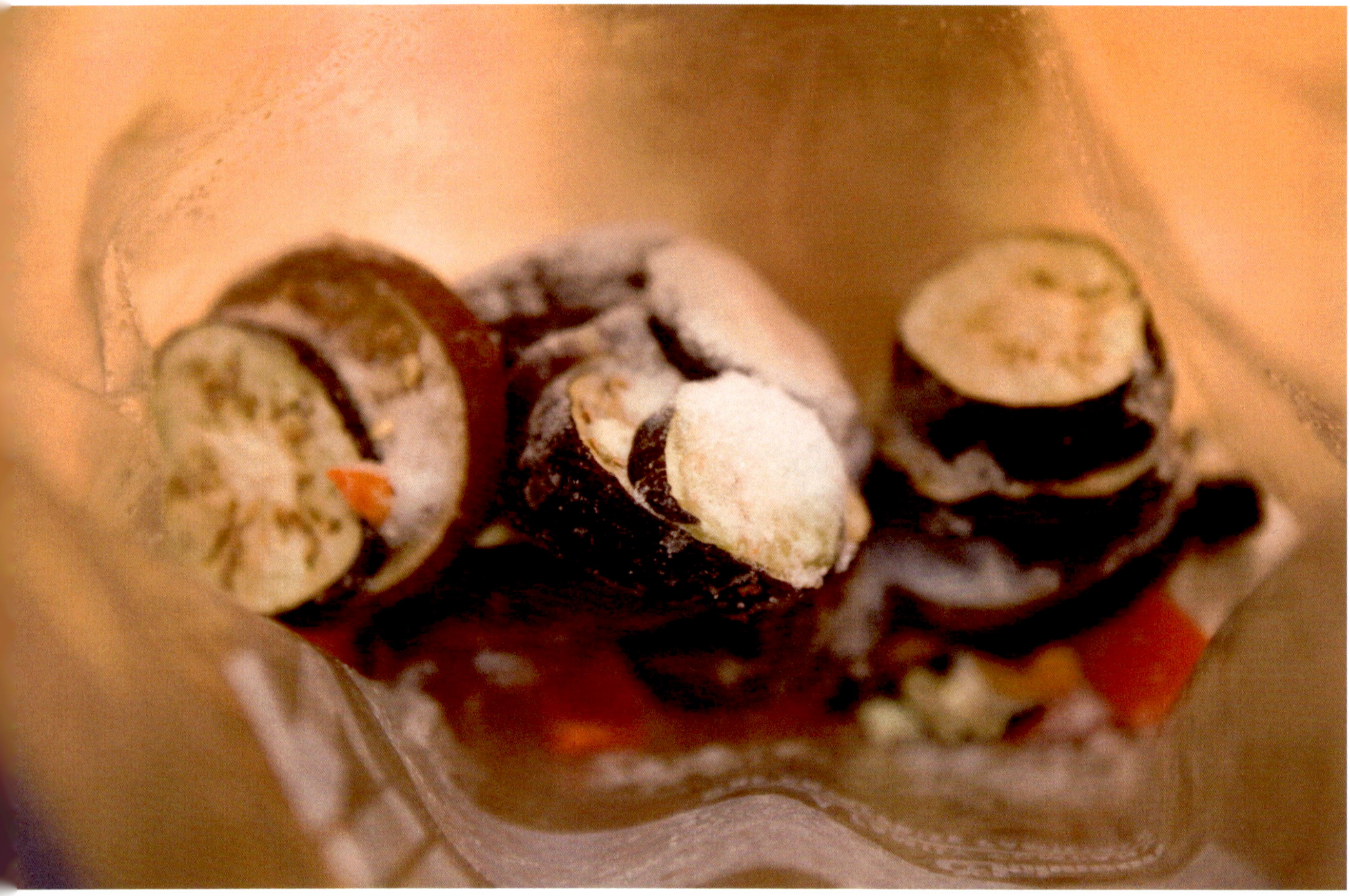

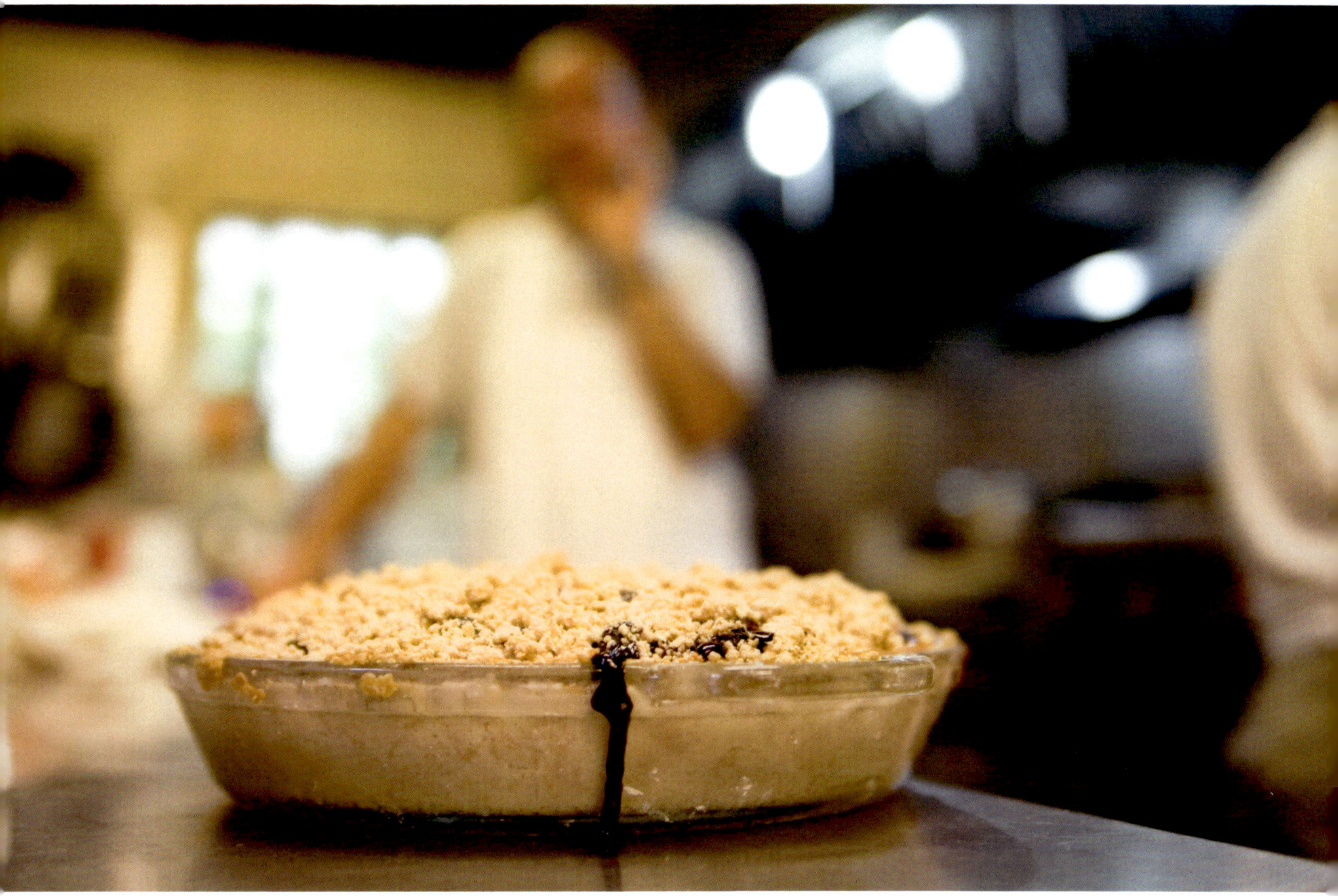

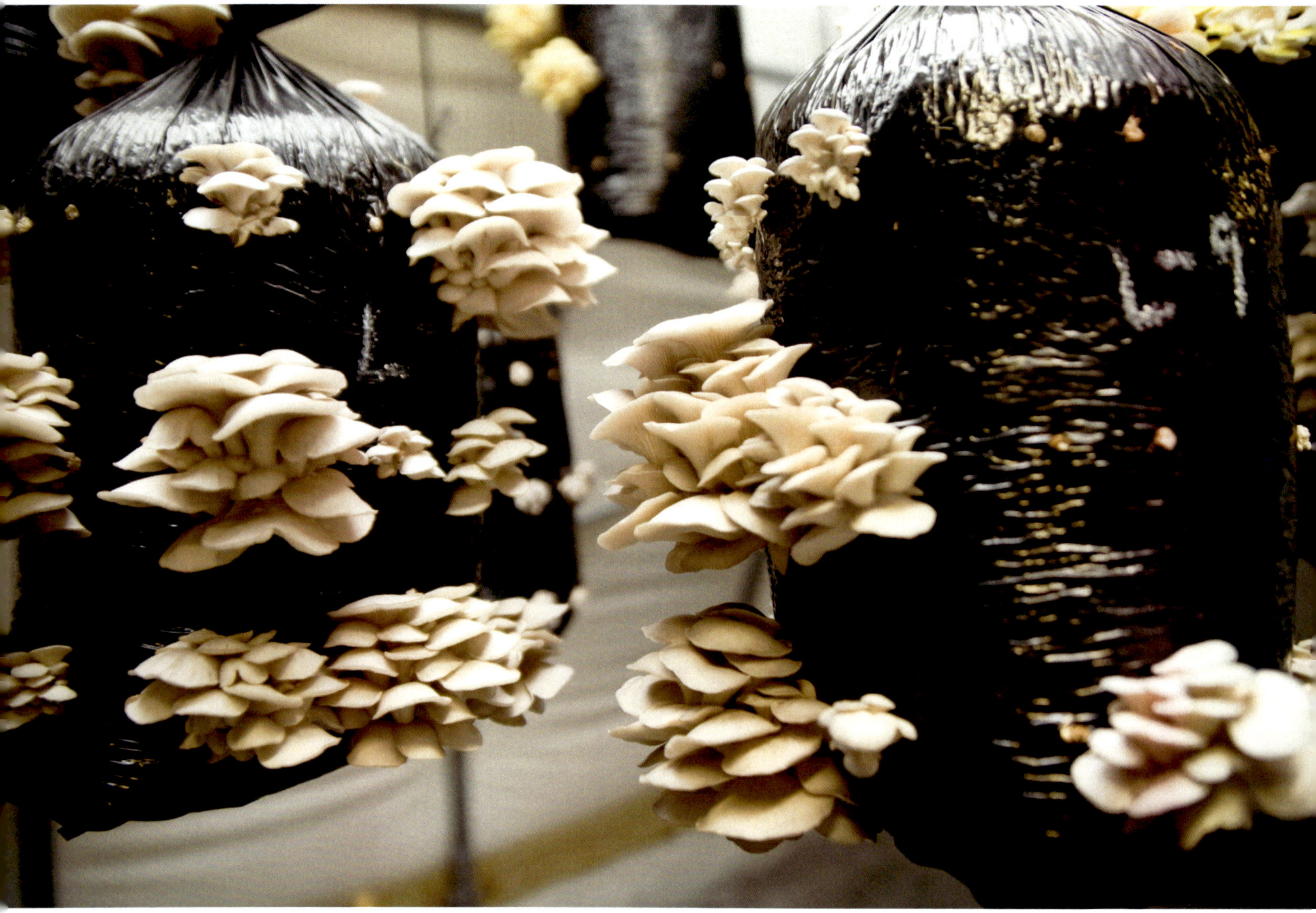

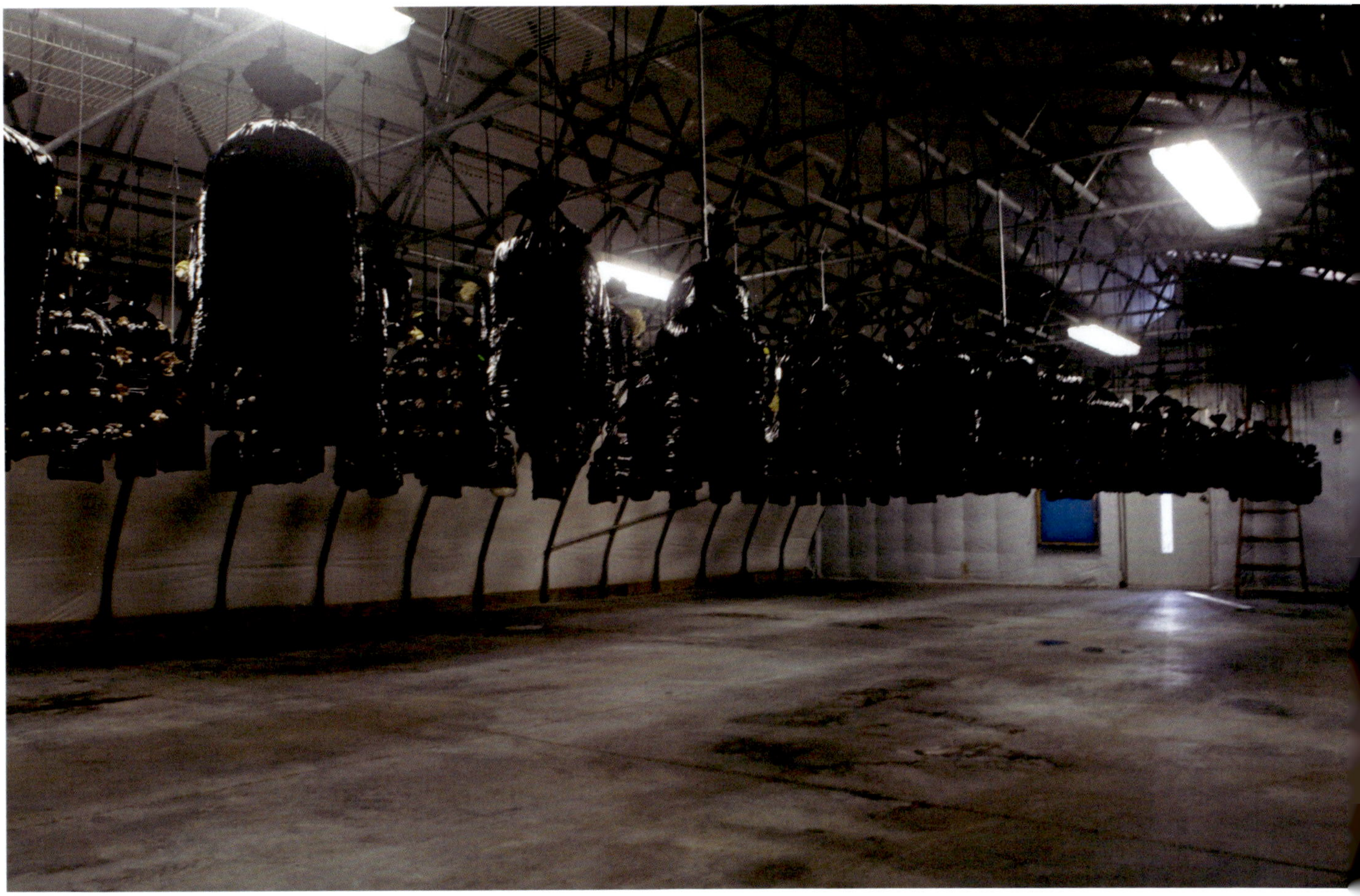

Dave and Dee's Homegrown Mushrooms produce several varieties of oyster mushrooms using grow bags in a highly controlled environment. The mushrooms only fruit at very specific temperatures and humidity levels—so precise that they can only produce one or two varieties at a time. The fruit takes 24 hours or less to grow and keeps Dave and Dee busy all year.

Eggplants go unpicked at Dayspring Farm because of a lack of labor as they concentrate on planting fall crops and harvesting vegetables just coming into season.

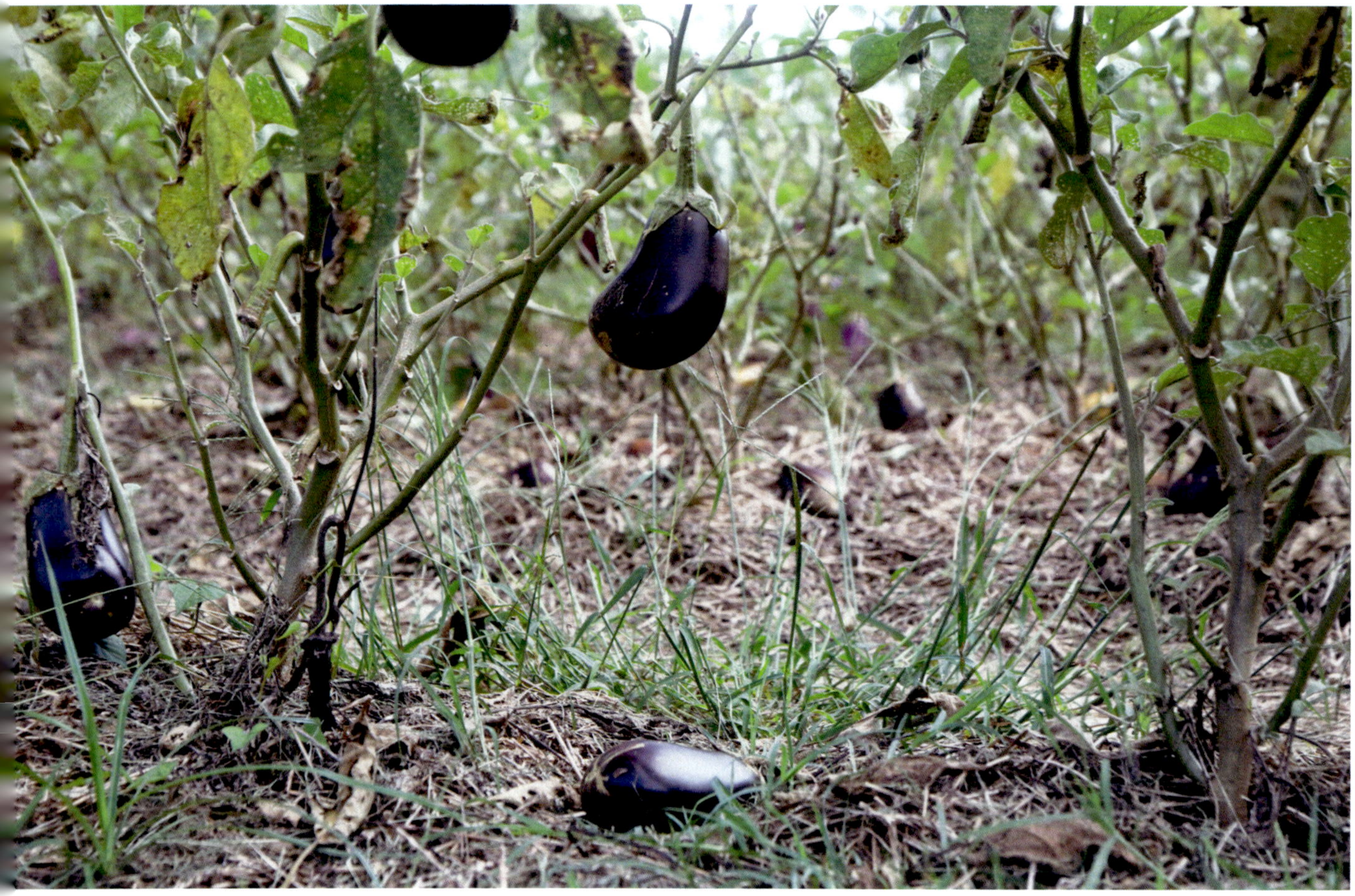

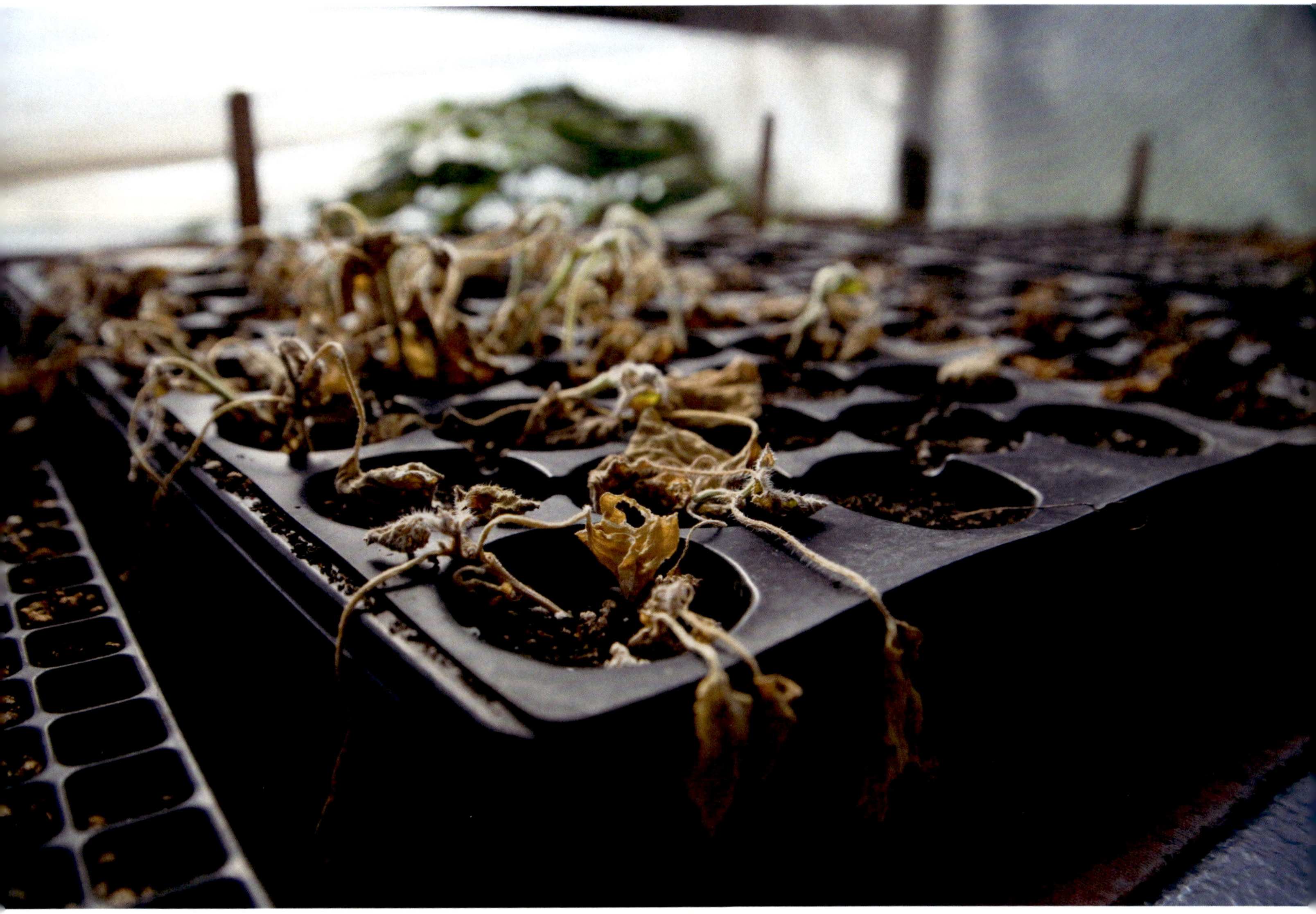

In the late summer, seedlings are neglected in the hoop house as farmers focus on maintaining their newly planted fall crops and harvesting the rest of their summer crops.

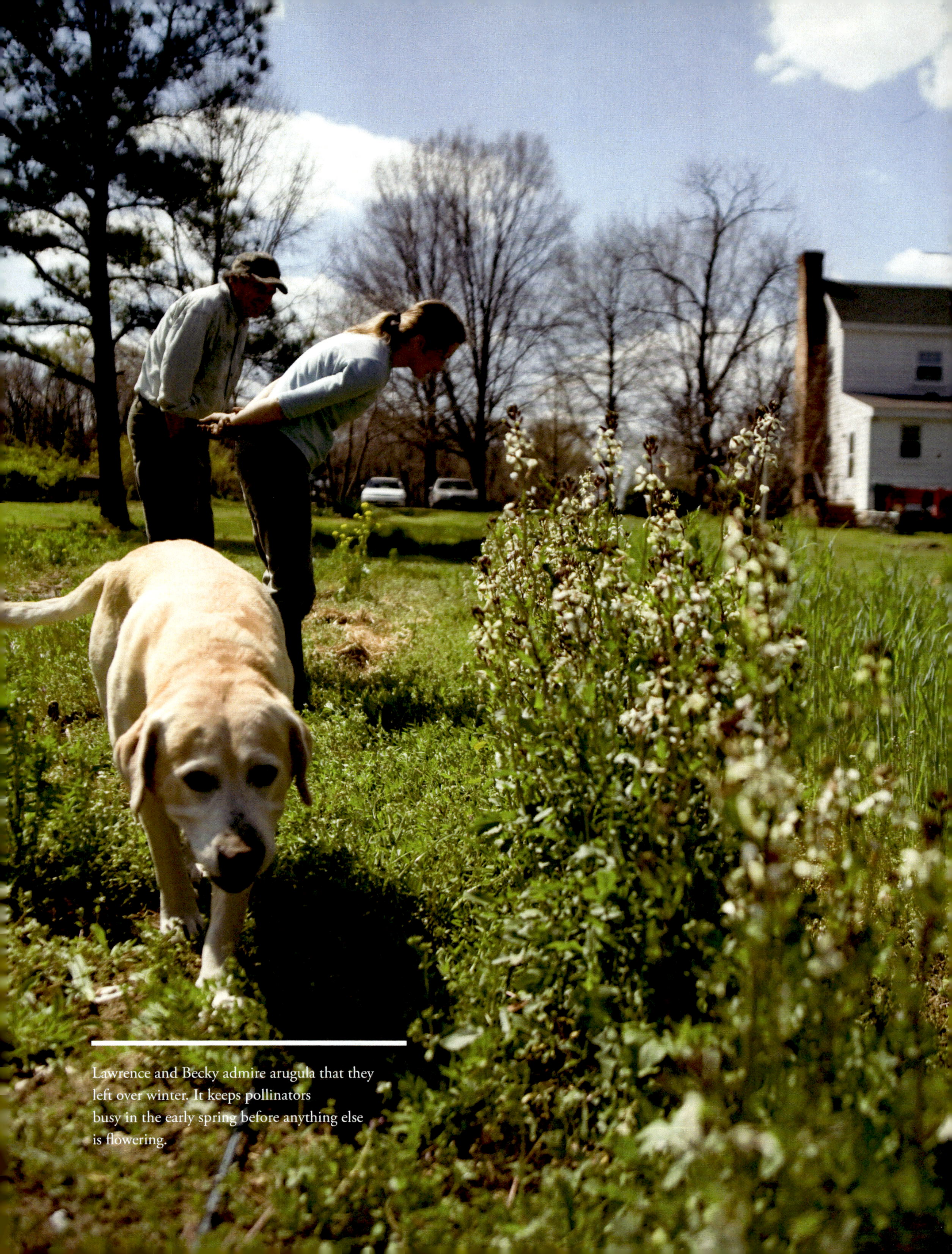

Lawrence and Becky admire arugula that they left over winter. It keeps pollinators busy in the early spring before anything else is flowering.

LAWRENCE LATANÉ

When Daniel Webster wrote, "the cultivation of the earth is the most important labor of man," I think he was thinking about the small farmer whose back is sore and neck is sunburned at the end of another endless day.

We've had the image of the yeoman farmer at least since the Bible. Lately, books, movies and newspaper articles have amplified the age-old message and presented those of us hardheaded enough to farm organically and sustainably in flattering ways.

They point out, we're the ones preserving "heirloom" seeds, conserving "biodiversity" and raising cows, sheep, pigs and chickens in the most "humane" and "ethical" ways.

Yet, why some of us go to such trouble is harder to explain than what it is we do.

My old journals don't offer any help. When I look through my notebooks I find entries as dry as the weather that frequently does its best to spoil my plans: "Cultivated garlic with Farmall Cub," read the single notation for March 3, 2007.

Two days later, my wife, Becky, penned a more revealing (and honest) line when she indicated we tried to plant peas in the woody scaffolding of the previous year's basil crop.

"Too hard and wet," she wrote.

My entry for April 18 that year didn't begin to express what I really must have said: "Overnight deer ate two full rows of chard by Blenheim House—400 plants." I'm sure I cussed the deer, cussed my luck, and wondered what in the world I was doing.

It's still a good question.

When I quit my day job in September 2008 to farm full time after eight years of part-time farming, I likened it to taking the first step backward from a belay off a cliff. I relied on trust and momentum to carry me down.

I have a better idea of what motivates Becky and me, now, but it seems naïve to talk about optimism. But, what else is sowing seeds this year in the same fields that baked under last year's drought and record heat?

Despite uncertainty, here we are again as I write this in February 2011, preparing for another year, another season, another crop, and another try at getting it right.

As organic farmers, we've adjusted to the long learning curve of agriculture. Mistakes one year usually can't be corrected until the next due to the calendar's relatively narrow limits of when crops can be planted and harvested. Even then, the solution depends on the farmer having a good enough memory to recall the blunder from 12 months earlier and still have the energy to correct the matter.

I wrote earlier about organic and sustainable farmers being hardheaded. There's a grain of truth to that in my case. I find a perverse pleasure in doing things the hard way. But, the longer I farm organically, the clearer I see it for what it is. It is the better way. In many ways, it's the easier way. It's certainly easier to let the soil do most of the heavy lifting after you've bolstered its fertility with cover crops.

Indeed, mankind has fed itself since time immemorial through the simple act of farmers tending their crops and flocks. It's not a question of if, but when the world's current experiment with substituting cheap oil for natural methods and technology for husbandry will literally run out of gas.

Studies in Pennsylvania already show that organically grown corn yields bigger crops than conventionally grown plots during drought, even as some of the big seed companies invest public-relations budgets to make us believe that genetically modified seeds are the only hope for what's apt to be a challenging future.

At every farmers market Becky and I attend, we hear firsthand how our certified organic vegetables are the best produce our customers have had.

But, the longer I farm organically, the clearer I see it for what it is. It is the better way. In many ways, it's the easier way.

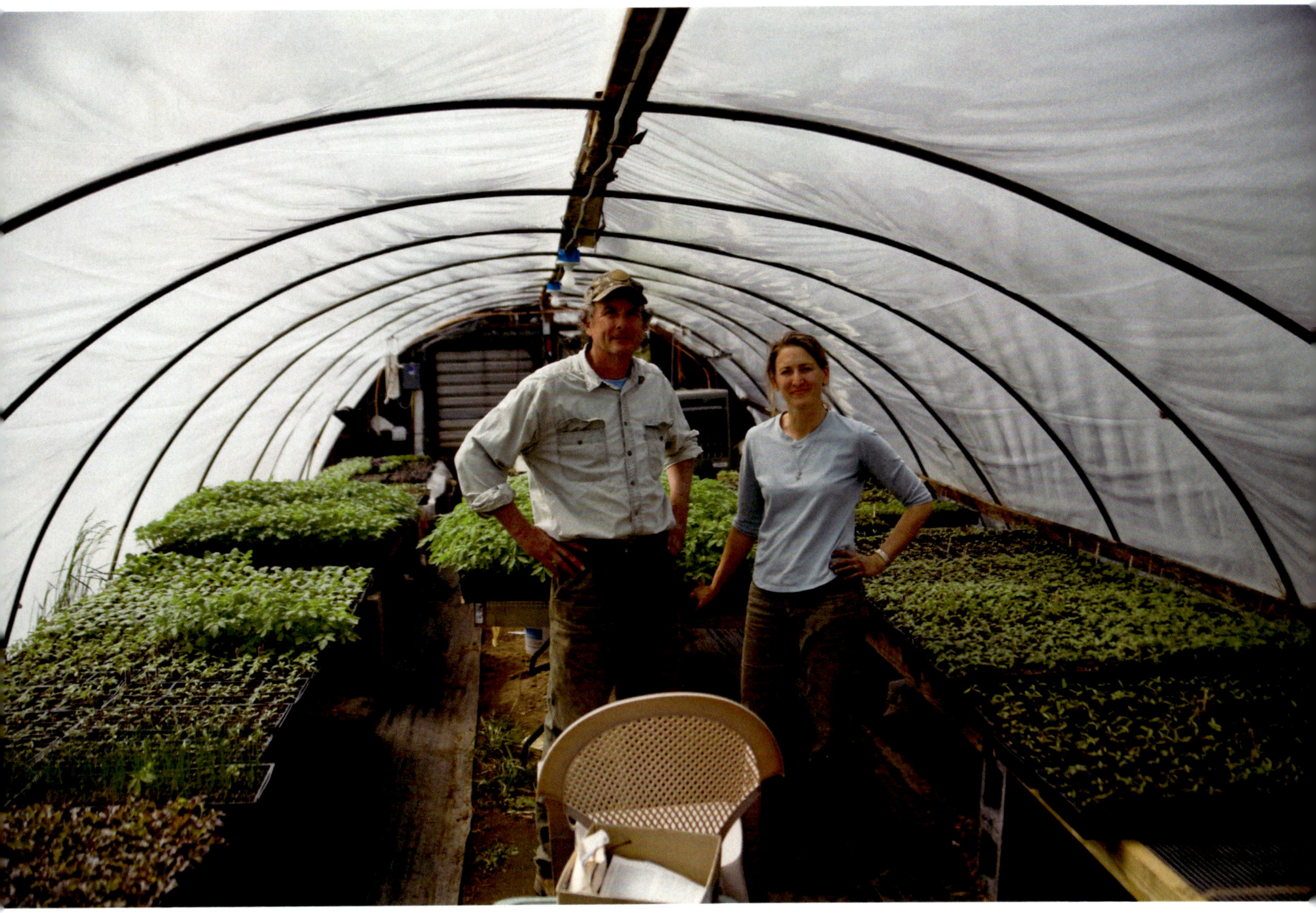

Lawrence and Becky in one of their hoop houses.

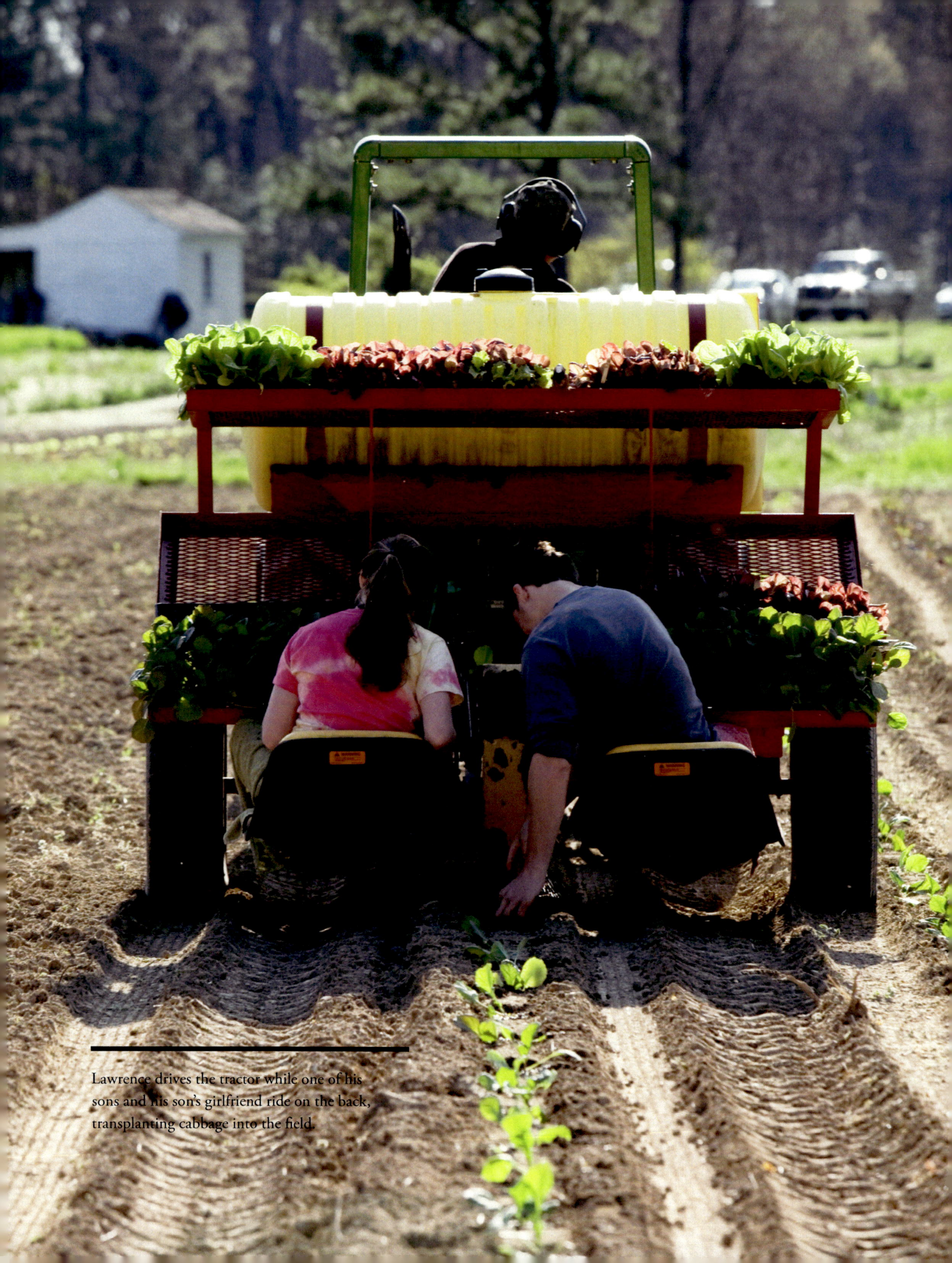

Lawrence drives the tractor while one of his sons and his son's girlfriend ride on the back, transplanting cabbage into the field.

Part of the reward of farming our way is this immediate feedback. The farmers market is the ultimate free-marketplace. No wonder that farmers practicing organic and sustainable methods are thriving worldwide.

When I got into this business, I was looking for a way for the farmland I lived on to pay its own way. In the process I traded a white collar job with a salary and benefits for a chance at trying my hand in the emerging green economy that didn't guarantee anything.

There are still no guarantees. But, every year we seem to do a little better, even though the challenges seem to increase.

Best of all, there's a covey of quail in my asparagus patch, goldfinches in the sunflowers, and honey bees buzzing in the clover. We might work 12-hour days, but our daily commute is exactly one step out the backdoor.

And, we live close to nature with few distractions to filter out its messages. Sometimes, even on our busiest days, there are transcendent moments not unlike the best moments that can be experienced on high mountain trails, trout streams or duck blinds.

Just last week, while cutting salad mix on a morning of off-and-on showers, I looked up and watched the puddles shimmer and come alive as fresh raindrops came down again.

I should have been miserable. The seat of my pants were wet and the seams of my rain jacket had failed. But the dogwoods along the field edge were starting to bloom, and the rye in the cover crop was growing tall and beginning to unfurl.

What a life, I thought, and bent back down to cut more greens.

Lawrence Latané operates Blenheim Organic Gardens with his wife Becky. The farm is certified organic and delivers produce to Fredericksburg and Northern Virginia.

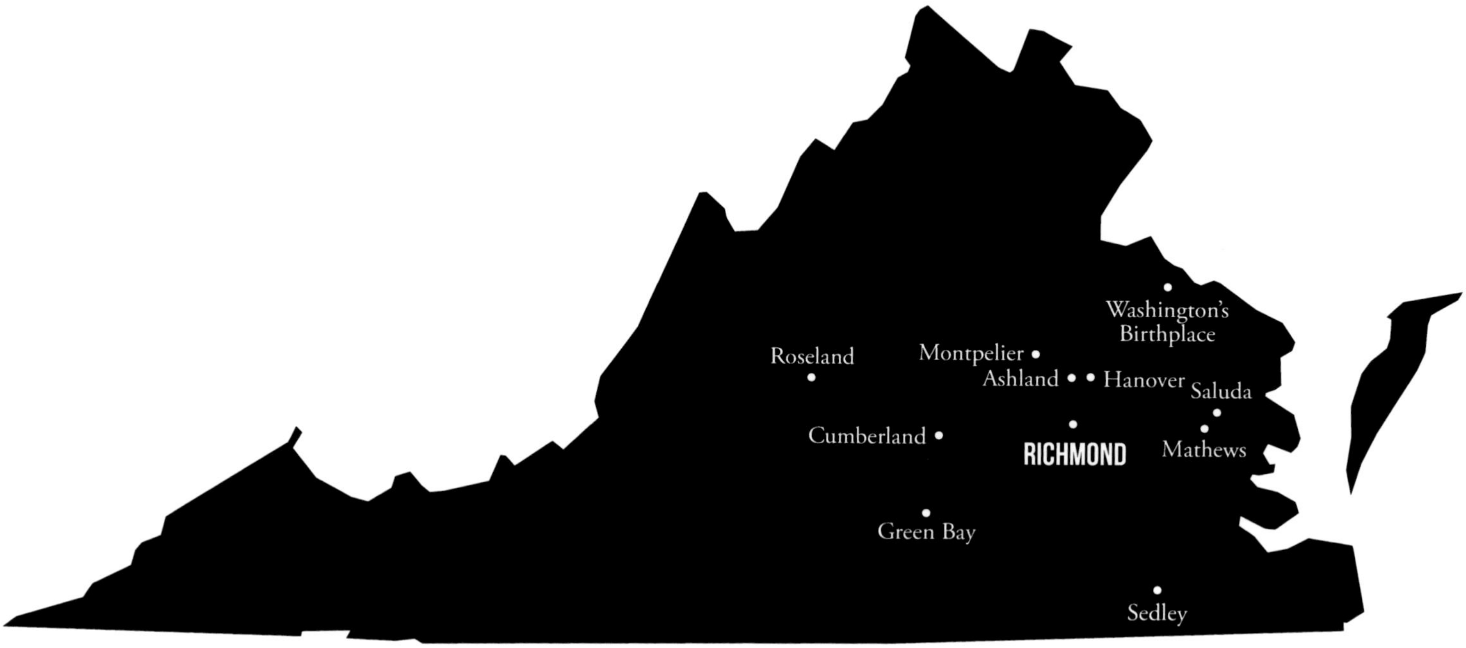
Washington's
Birthplace
Roseland
Montpelier
Ashland
Hanover
Saluda
Cumberland
RICHMOND
Mathews
Green Bay
Sedley

FEATURED IN THE PROJECT

17th Street Garden
Richmond

Agriberry CSA
Hanover County

Appalachia Star Farm
Roseland

Ashland Farmers Market
Ashland

Bill and Kim
Richmond

Blenheim Organic Gardens
Washington's Birthplace

Boom Boom Burgers
Richmond

Byrd House Market
Richmond

Café Gutenburg
Richmond

Dave and Dee's
Homegrown Mushrooms
Sedley

Dayspring Farm
Cologne

Dragon Creek Farm
Saluda

Eastfield Farms
Mathews

Edible Gardens
Richmond

Faith Farm
Green Bay

Farm To Family
Richmond

Fertile Crescent
Green Bay

Green Bay Market
Green Bay

Mamma Zu's Garden
Richmond

Red Dog Farm
Hanover County

Rachel and Lee
Richmond

Simply Abundant Farm
Richmond

Sub Rosa Bread
Richmond

Sweet Tree Hill
Cumberland

Susannah and Chandler
Richmond

Trail's End Farm
Montpelier

Tricycle Gardens
Richmond

- 9th + Bainbridge
- Chimborazo Garden
- Jefferson Garden

University of Richmond
Community Garden
Richmond

Victory Farms
Hanover County

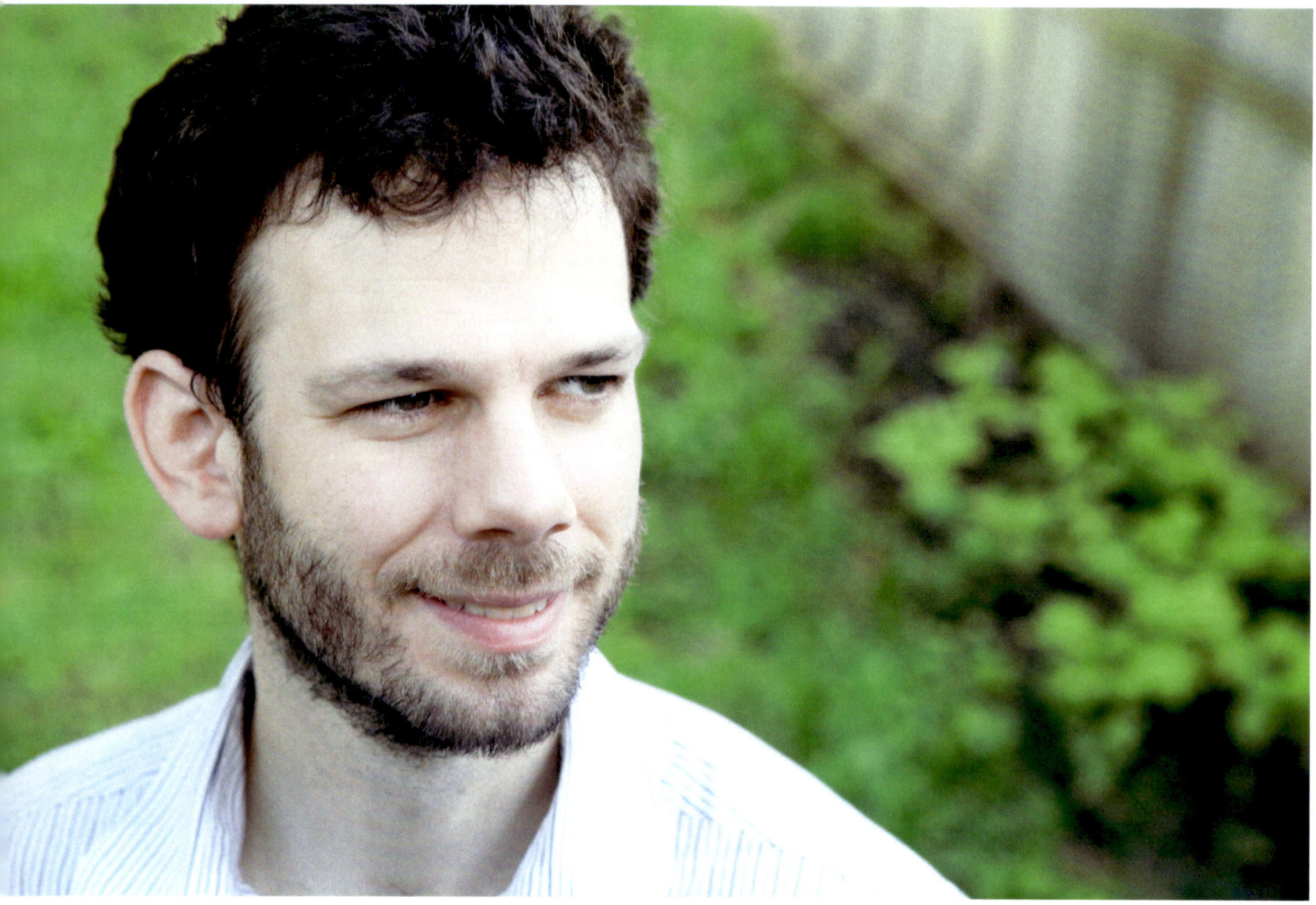

ABOUT THE PHOTOGRAPHER

From his roots in Poughkeepsie, New York to his professional life in Richmond, Virginia, Gordon Stillman has explored the world through photography. He got his start with photography classes in his New Hampshire high school. While majoring in environmental studies and philosophy at the College of William and Mary, he began to explore these disciplines through photography. Stillman's work focuses on exploring human relationships with their environments to gain insight into the world and understand how societies can improve experiences and existence in their environments.

For more on Gordon Stillman's work, visit www.gordonstillman.com.

WINTER

SPRING

SUMMER

Page 31
Drying Garlic
Fertile Crescent Farm
Green Bay

Page 32
Turkey
Simply Abundant Farm
Richmond

Page 34
Pollinating Arugula
Blenheim Organic Gardens
Washington's Birthplace

Page 35
Honeycomb
Blenheim Organic Gardens
Washington's Birthplace

Page 36
Chicken Wing Clippings
Faith Farm
Green Bay

Page 37
Escaped Chicken
Simply Abundant Farm
Richmond

Page 38
Pasture
Faith Farm
Green Bay

Page 40
Cover Crop
Red Dog Farm
Hanover County

Page 41
Onion Blossom
Red Dog Farm
Hanover County

Page 42
Gathering Eggs
Trail's End Farm
Montpelier

Page 43
Flowers at Fertile Crescent
Fertile Crescent Farm
Green Bay

Page 44
Grazing Cows
Faith Farm
Green Bay

Page 45
Pigs in the Field
Faith Farm
Green Bay

Page 47
Curious Pigs
Faith Farm
Green Bay

Page 48
Pamunkey River
Agriberry
Hanover County

Page 49
Open Drip Irrigation
Agriberry
Hanover County

Page 50
Flock of Sheep
Simply Abundant Farm
Richmond

Page 51
Watering Pails
Simply Abundant Farm
Richmond

//////////////////////////////////

SUMMER (CONTINUED)

Page 53
Shorn and Unshorn Sheep
Faith Farm
Green Bay

Page 54
Green Tomato
Victory Farms
Hanover County

Page 55
Arugula
Fertile Crescent Farm
Green Bay

Page 56
Red Dog
Ashland Farmers Market
Ashland

Page 57
CSA Pickup at Byrd House Market
Byrd House Market
Richmond

Page 59
Garden in the Bottom
17th Street Garden
Richmond

Page 60
Kim and Ewan Selecting Tomatoes
Kim and Ewan at Victory CSA pickup, Bryan Park Market
Richmond

Page 61
Garlic Creamed Chard
Victory Farms Potluck
Hanover County

Page 62
Okra at Mamma Zu's
Mamma Zu's Garden
Richmond

Page 63
Sub Rosa Loaves
Sub Rosa Bread
Richmond

Page 64
Mobile Food Market
Farm To Family
Richmond

Page 66
Preparing Local Chickens
Edible Gardens restaurant
Richmond

Page 67
Fried Green Tomato
Café Gutenberg
Richmond

Page 68
Lisa Taranto
Tricycle Gardens
Richmond

Page 69
Jefferson Garden
Tricycle Gardens
Richmond

Page 70
Building Chimbo Community Garden
Tricycle Gardens
Richmond

FALL

Page 71
Chimborazo Garden Beds
Tricycle Gardens
Richmond

Page 72
Feeding the Pigs
Faith Farm
Green Bay

Page 74
Cantaloupe
Trail's End Farm
Montpelier

Page 75
Squash
Dayspring Farm
Cologne

Page 76
Carolyn
Green Bay Open Air Market
Green Bay

Page 77
Market Offerings
Green Bay Open Air Market
Green Bay

Page 78
Jam Jars
Agriberry at the
Hanover County Cannery
Hanover County

Page 79
Anne Cooking Jam
Agriberry at the
Hanover County Cannery
Hanover County

Page 80
Thawing Eggplant
Susannah and Chandler's
Richmond

Page 81
Berry Pie
Edible Garden's Restaurant
Richmond

Page 82
Oyster Mushrooms
Dave and Dee's
Homegrown Mushrooms
Sedley

Page 83
Growing Bags
Dave and Dee's
Homegrown Mushrooms
Sedley

Page 84
Fallen Eggplant
Dayspring Farm
Cologne

Page 85
Neglected Seedlings
Fertile Crescent Farm
Green Bay

////////////////////////////////

MISCELLANEOUS

Cover
Greens
Victory Farms
Hanover County

Page 6
Kim Gardening
University of Richmond
Community Garden
Richmond

Page 9
Corn Stalks from Last Season
Appalachia Star Farm
Roseland

Page 10
Adam and Lucy
Fertile Crescent Farm
Green Bay

Page 74
Gordon Stillman
Courtesy of Robert Elves

Page 86
Flowering Arugula
Blenheim Organic Gardens
Washington's Birthplace

Page 89
Lawrence and Becky
Blenheim Organic Gardens
Washington's Birthplace

Page 90
Transplanting Cabbage
Blenheim Organic Gardens
Washington's Birthplace